Infosys Science Foundation Series

Infosys Science Foundation Series in Mathematical Sciences

The *Infosys Science Foundation Series in Mathematical Sciences* is a sub-series of The *Infosys Science Foundation Series*. This sub-series focuses on high quality content in the domain of mathematical sciences and various disciplines of mathematics, statistics, bio-mathematics, financial mathematics, applied mathematics, operations research, applies statistics and computer science. All content published in the sub-series are written, edited, or vetted by the laureates or jury members of the Infosys Prize. With the Series, Springer and the Infosys Science Foundation hope to provide readers with monographs, handbooks, professional books and textbooks of the highest academic quality on current topics in relevant disciplines. Literature in this sub-series will appeal to a wide audience of researchers, students, educators, and professionals across mathematics, applied mathematics, statistics and computer science disciplines.

More information about this series at http://www.springer.com/series/13817

Phoolan Prasad

Propagation of Multidimensional Nonlinear Waves and Kinematical Conservation Laws

Springer

Phoolan Prasad
Department of Mathematics
Indian Institute of Science
Bengaluru, Karnataka
India

ISSN 2363-6149 ISSN 2363-6157 (electronic)
Infosys Science Foundation Series
ISSN 2364-4036 ISSN 2364-4044 (electronic)
Infosys Science Foundation Series in Mathematical Sciences
ISBN 978-981-13-3970-7 ISBN 978-981-10-7581-0 (eBook)
https://doi.org/10.1007/978-981-10-7581-0

Printed on acid-free paper

This Springer imprint is published by Springer Nature
The registered company is Springer Nature Singapore Pte Ltd.
The registered company address is: 152 Beach Road, #21-01/04 Gateway East, Singapore 189721, Singapore

Preface

In this monograph, we deal with a method to trace the evolution of a surface. Such an evolution is a very common phenomenon in nature—from the surface of a soap bubble and that of a growing crystal to the propagation of a wavefront in cosmology. In all such phenomena, there is a need to find successive positions of a moving curve Ω_t in a plane or surface Ω_t in space from a given initial position Ω_0. One conspicuous feature is that Ω_t may develop singularities on it even if Ω_0 is smooth. Then, the evolution equations of Ω_t, if written in the form of differential equations, are no longer valid, and we need to go back to the basic conservation laws. Though the differential form of the equations was known for a long time as *ray equations*, the basic conservation laws for the evolution of a curve in a plane were discovered in 1992 by Morton, Prasad and Ravindran and were called 2-D *Kinematical Conservation Laws* (KCLs). The conservation form of evolution equations for a surface in 3-D space, called 3-D KCL, was discovered by Giles, Prasad and Ravindran in 1995. The main aim of this monograph is to derive KCL, analyse them and present their applications to some problems in physics.

There are some other methods to deal with the evolution of curves and surfaces, but KCL contains an intrinsic property to account for the evolution of a surface with singularities. No other approach takes into account these physically realistic singularities in a very natural way. The proposed material in this book is an outcome of collaborative work carried out mainly at the Indian Institute of Science, Bengaluru. Chapter 1 contains a summary of the whole material in this book.

To make the book self-contained, we wish to include in it the theory of hyperbolic partial differential equations and conservation laws very briefly. Though these are highly developed and hot topics of research today (see only two books [15] and [47]; both contain a very large number of references), we give only a simple presentation of these two topics needed for our purpose. Chapter 2 contains first-order partial differential equations. Using a single conservation law in two independent variables as a model, we develop the concept of shocks in Chap. 3. The theory of a hyperbolic system of n equations in multi-dimensions (i.e. in more than two independent variables) and a discussion of shocks in multi-space dimensions are generally not available in most books, and hence, we include a brief introduction to this topic in Chap. 4, starting first with a discussion of the system in two independent variables. The notions developed in these three chapters play an important role in the theory of KCL and applications. Chapter 5 contains a brief derivation of the equations of weakly nonlinear ray theory and weak shock propagation. They represent equations of physically realistic phenomena, with the help of which KCL equations can be fully analysed and used to solve some practical problems. Rest of the book is devoted to the derivation of KCL systems, their analysis and application to some problems in physics.

It is recommended that every reader goes through the introduction to the subject in Chap. 2. It not only introduces the subject and lists the main equations but also summarizes the material in the whole book, I believe, in a simple way. *This book is a research note and not a research monograph or a textbook. Hence, the material presented in this book is in the form of an extended summary of some results and a reader needs to look into other books and research papers, cited in the bibliography, if he desires to know some details.*

Two of my former students, who have contributed significantly to the development of KCL theory and its applications, have also helped me in the preparation of this book. They are S. Baskar (he joined me in working on many diverse applications of 2-KCL) and K. R. Arun (he joined me in completing difficult analysis of 3-D KCL, developed a numerical scheme for it and also used it to solve some problems).

The National Academy of Sciences, India, has been generous to extend my Senior Scientist Platinum Jubilee Fellowship, the last two years of which were spent mainly in writing this book. The Department of Mathematics, Indian Institute of Science, Bengaluru, where I have worked since 1965, continues to provide me with an office space and facilities for all academic work. It has been a great privilege for me to have been in good contact with all the existing members of this department even after my superannuation.

It is more than 20 years since our children Deepika and Amritanshu went away from home for higher studies and their jobs. Since then, my wife Mandra and I have been living alone. Even after my formal retirement in 2006, I continued to have some academic positions, which demanded an intense involvement in work. During the last three years, the pressure of the work decreased and it was expected that I should have spent more relaxed time at home. However, in addition to organizing and participating in lecture workshops for teachers and students all over India, I took up writing of this book. I thank Mandra for providing support to my academic work, especially during the period of preparation of this book, in the same way, she had done since 1969.

Bengaluru, India
October 2017

Phoolan Prasad

Contents

About the Author

Phoolan Prasad worked in the Department of Mathematics, Indian Institute of Science (IISc), Bengaluru, India. With over 52 years of experience, he is an Indian mathematician who specialized in partial differential equations and fluid mechanics. He held a distinguished chair of Mysore Sales International Limited (MSIL) Professorship and was the Chairman of the Department and Professor in super-time scale before his retirement. Later, he was Honorary Professor and Department of Atomic Energy Raja Ramanna Fellow (DAE-RRF), Indian National Science Academy (INSA) Senior Scientist and The National Academy of Sciences, India (NASI)-Senior Scientist Platinum Jubilee Fellow. For his research contribution in nonlinear waves, Government of India awarded him Shanti Swarup Bhatnagar Prize in 1983. He is Fellow of all science academies in India: NASI; Indian Academy of Sciences (IASc) and INSA.

He edited "Nonlinear Waves in One-dimensional Dispersive Systems" by P.L. Bhatnagar, Oxford University Press (1979), and jointly authored a book "Partial Differential Equations", John Wiley and Wiley Eastern (1984). Most of the material of his books "Propagation of a curved shock and Nonlinear Ray Theory" by Longman Higher Education (1993) and "Nonlinear Hyperbolic Waves in Multi-dimensions", Chapman & Hall/CRC (2001) is based on the research work of his group.

With a deep interest in education, he has been playing an important role in the development of quality education in mathematics from school education to research, e.g. International Mathematical Olympiad, lecturing in and organizing numerous workshops, writing articles for the students and teachers and taking initiative in planning and implementing new activities all over India.

In this book, he gives a sketch of nonlinear ray theory developed since 1975 and of shock ray theory developed since 1982. The theory of kinematical conservation laws, developed since 1992 by his research group, forms the theme of this book.

About the Author

Chapter 1
Introduction

The running theme in this monograph is **hyperbolic nature** of a system of partial differential equations (PDEs), which is a system in which one independent variable plays a distinctive (or special) role. We denote this **time-like** variable by t and other **spatial** variables by $\boldsymbol{x} = (x_1, x_2, \ldots, x_d) \in \mathbb{R}^d$. The hyperbolic nature of the system is due to the fact that the system has a sufficient number (or full set) of families of curves in the space–time which carry information from the initial plane $t = 0$ to solve an initial value problem (*a Cauchy problem*). For two independent variables ($d = 1$), these curves are **characteristic** curves and for more than two independent variables ($d > 1$), they are **bicharacteristic** curves (which can be identified with rays in physical $\boldsymbol{x}$-space). We shall show that, in this sense, first-order PDEs are the simplest examples of hyperbolic equations.

The basic materials which are important for some deeper understanding of hyperbolic equations and which are needed for **kinematical conservation laws (KCL)** do not require analysis in the form of too many proofs but only require study of the structure of the equations and a clear derivation of results. We aim that this monograph should be useful to scientists working in areas of applications and to mathematicians to learn a new system of equations (namely KCLs) and their properties and to develop it further. Keeping above two aims in mind, we have left out many proofs, which can be found in the books [13, 17, 51].

In Chaps. 2, 3 and 4, we shall cover the theory of the following four PDEs briefly:

- General form of a first-order partial differential equation for $u : \mathbb{R}^{d+1} \rightarrow \mathbb{R}$ in $d + 1$ independent variables $\boldsymbol{x}, t$

$$F(\boldsymbol{x}, t, u, \nabla_{\boldsymbol{x}} u, u_t) = 0. \tag{1.1}$$

- A particular case of a scalar (or single) conservation law for $u : \mathbb{R}^2 \rightarrow \mathbb{R}$ in 2 in two independent variables x and t

P. Prasad, *Propagation of Multidimensional Nonlinear Waves and Kinematical Conservation Laws*, Infosys Science Foundation Series,
https://doi.org/10.1007/978-981-10-7581-0_1

$$u_t + \left(\frac{1}{2}u^2\right)_x = 0. \tag{1.2}$$

- A system of n first-order quasilinear PDEs in two independent variables x and t

$$A(x,t,\ \boldsymbol{u})\,\boldsymbol{u}_t + B(x,t,\ \boldsymbol{u})\boldsymbol{u}_x + \boldsymbol{C}(x,t,\ \boldsymbol{u}) = 0, \tag{1.3}$$

where we assume that the matrix A is nonsingular, and
- A system of n first-order quasilinear equations in $d+1$ independent variables $\boldsymbol{x}$ and t

$$A(\boldsymbol{x},t,\ \boldsymbol{u}))\,\boldsymbol{u}_t + B^{(\alpha)}(\boldsymbol{x},t,\ \boldsymbol{u})\,\boldsymbol{u}_{x_\alpha} + \ \boldsymbol{C}(\boldsymbol{x},t,\ \boldsymbol{u}) = 0, \tag{1.4}$$

where $\boldsymbol{u} \in \mathbb{R}^n$, $A \in \mathbb{R}^{n\times n}$, $B \in \mathbb{R}^{n\times n}$, $B^{(\alpha)} \in \mathbb{R}^{n\times n}$ and $\boldsymbol{C} \in \mathbb{R}^n$.

The theory of (1.1) in two and that in more ($d > 1$) independent variables are based on the same principle, derivation of characteristic equations and compatibility conditions, but we shall have to deal with its quasilinear and fully nonlinear cases separately (see Sects. 2.1 and 2.3). The theory of hyperbolic systems in two and $d+1$-dimensions with $d > 1$ are quite different—it is here that the distinction between time-like variable and space-like manifold becomes really important [13, 42, 47].

Definition 1.0.1 Given a function $\boldsymbol{u}_0(\boldsymbol{x})$ (or $u_0(\boldsymbol{x})$ for (1.1)) defined in a domain in $\mathbb{R}^d$, **Cauchy initial value problem** consists in finding a solution[1] $\boldsymbol{u}(\boldsymbol{x},t)$ (or $u(\boldsymbol{x},t)$ for (1.1)) of the above equations in a domain in $(\boldsymbol{x},t)$-space such that

$$\boldsymbol{u}(\boldsymbol{x},0) = \ \boldsymbol{u}_0(\boldsymbol{x}). \tag{1.5}$$

The above equations deal with phenomena, in which a solution $\boldsymbol{u}$ is smooth, but we shall see later, $\boldsymbol{u}$ is quite often discontinuous on surfaces in the space of independent variables. Then, quite often we need to replace the differential equations by their conservation forms from which they are derived. Suppose the matrices A and $B^{(\alpha)}$ are functions of $\boldsymbol{u}$ only and are such that there exist vector $\boldsymbol{H} \in \mathbb{R}^n$ and a matrix $\tilde{\boldsymbol{F}} = (\boldsymbol{F}^{(1)}, \boldsymbol{F}^{(2)}, \ldots, \boldsymbol{F}^{(d)}) \in \mathbb{R}^{n\times d}$ satisfying

$$A(\boldsymbol{u}) = \langle \nabla_{\boldsymbol{u}}, \boldsymbol{H}\rangle \ \ \textit{and} \ \ B^{(\alpha)}(\boldsymbol{u}) = \langle \nabla_{\boldsymbol{u}}, \boldsymbol{F}^{(\alpha)}\rangle. \tag{1.6}$$

Now, for a smooth $\boldsymbol{u}$, the system (1.4) with $\boldsymbol{C} = 0$ can be derived from a system of *conservation laws*[2] in $d+1$ independent variables

$$\boldsymbol{H}_t(\boldsymbol{u}) + \langle \nabla_{\boldsymbol{x}}, \tilde{\boldsymbol{F}}(\boldsymbol{u})\rangle = 0 \tag{1.7}$$

[1] In this book, we shall deal with two types of solutions: **genuine solution** of (1.1), (1.3) or (1.4) which is a $\mathbb{C}^1$ function satisfying the equations pointwise and **weak solution** of (1.2) or (1.7) which we shall define in Chap. 3.

[2] The system (1.7) is just a symbolic representation of a system of conservation laws written in integral form, which we shall explain later in Sect. 3.1.

which we write more explicitly as

$$H_t(u) + F^{(1)}_{x_1}(u) + F^{(2)}_{x_2}(u) + \ldots + F^{(d)}_{x_d}(u) = 0. \tag{1.8}$$

Consider a surface Ω_t, which evolves with time in $\boldsymbol{x} = (x_1, x_2, \ldots, x_d)$-space (i.e. $\mathbb{R}^d$). The d-dimensional kinematical conservation laws (KCL) is a system of conservation laws governing the evolution of the surface Ω_t in $\mathbb{R}^d$ and is derived in a specially defined ray coordinates $(\boldsymbol{\xi}, t)$, where $\boldsymbol{\xi} = (\xi_1, \xi_2, \ldots, \xi_{d-1})$ are the surface coordinates on Ω_t and t is time. The mapping between the ray coordinates $(\boldsymbol{\xi}, t)$ and the spatial coordinates $\boldsymbol{x}$ is assumed to be locally one-to-one. Since the KCL is a system of conservation laws, its solutions may contain a $(d-1)$-dimensional shock manifolds in the d-dimensional ray coordinates. A $(d-2)$-dimensional **shock front** is the projection on $\boldsymbol{\xi}$-space of the section of a shock manifold by $t = constant$. Image of a shock, when mapped onto the $\boldsymbol{x}$-space, is a $(d-2)$-dimensional **kink**[3] on Ω_t across which the normal direction to Ω_t and its normal velocity are discontinuous. Hence, the KCL is ideally suited to study the evolution of a surface Ω_t having kink type of singularities on it. However, the KCL being purely a geometric result, it forms an incomplete system of equations and additional closure relations from the dynamics of the medium are necessary to get a completely determined set of equations to study the evolution of Ω_t. Derivation and discussion of some properties of d-D KCL without the closure relations is presented in Chap. 6.

When Ω_t is a **weakly nonlinear wavefront**[4] in a polytropic gas, the KCL system is closed by a single equation representing the conservation of total energy in a ray tube. We shall refer to the complete system of conservation laws governing the evolution of a nonlinear wavefront, as KCL-based weakly nonlinear ray theory (WNLRT), or briefly d-D WNLRT for evolution of the front in d-space dimensions.

For motion of a curve Ω_t in a plane ($d = 2$), the equations for 2-D WNLRT consist of just three conservation laws in (ξ, t) coordinates:

$$(gn_2)_t + (mn_1)_\xi = 0, \quad (gn_1)_t - (mn_2)_\xi = 0. \tag{1.9}$$

$$\{g(m-1)^2 e^{2(m-1)}\}_t = 0, \tag{1.10}$$

where m is an appropriately defined non-dimensional front velocity, $\boldsymbol{n} = (n_1, n_2)$ is unit normal to Ω_t and g is the metric associated with the variable ξ (here ξ_1 is replaced by ξ), i.e. $gd\xi$ is an element of length along Ω_t. This system of 2-D WNLRT equations is hyperbolic when the appropriately defined non-dimensional front velocity say, $m > 1$. In the case of a polytropic gas, we shall show that $m > 1$ corresponds to a wavefront on which the pressure is greater than the constant pressure

[3] We shall define a kink precisely in Sect. 6.3.

[4] We need to define a wavefront. See Sect. 4.1 for definition using the language of physics. A mathematical definition is available in [42], Sect. 3.2.1, originally given in a lecture note in 1977 [36].

in the ambient gas. In Chaps. 8 and 9 we shall discuss various properties of 2-D KCL, and present applications of 2-D WNLRT to some physical problems.

The simplicity of 2-D KCL is lost when we consider the 3-D KCL, which is a system of six conservation laws in (ξ_1, ξ_2, t) coordinates with three *stationary* divergence-free constraints. We shall refer to the three constraints together as **geometric solenoidal constraint**. These equations, with symbols of dependent variables and their definitions, are too complex to write in this introduction. When we add to the 3-D KCL the energy transport equation for a weakly nonlinear wavefront in 3-D, we get a system of seven conservation laws. This system of 3-D WNLRT equations has two distinct non-zero eigenvalues and an eigenvalue *zero* of multiplicity five with a 4-D eigenspace. When $m < 1$ the distinct eigenvalues are imaginary and when $m > 1$, they are real. Thus, when $m > 1$ all eigenvalues of 3-D WNLRT are real but the system is **weakly hyperbolic**. The weakly hyperbolic nature of the system and the presence of geometric solenoidal constraint pose a challenge to develop a numerical approximation. It is well known that, in general, the solution of a Cauchy problem for a weakly hyperbolic system contain **Jordan modes**, which grow algebraically in time. However, when the geometric solenoidal constraint is satisfied initially, the solution to a Cauchy problem does not exhibit the Jordan mode. Motivated by this, a **constraint transport (CT)** technique has been built into a central finite volume scheme for the 3-D WNLRT system, therein the constraint is maintained up to machine accuracy with elimination of the Jordan modes [1]. The numerical method is robust, second-order accurate and the numerical solution can be continued for a very long time, almost indefinitely. The theory of 3-D KCL and some applications have been presented in Chap. 10.

Though the use of KCL to study weakly nonlinear wavefronts gives a very deep understanding of the geometrical features of Ω_t, the ultimate problem in gas dynamics is to study propagation of curved shock fronts. Therefore, in Chaps. 9 and 10, we deal with the one of the most challenging problems in fluid dynamics: finding successive positions of a curved shock front Ω_t [5]. The challenge is in computing a shock front, (i) with very sharp geometrical shape of Ω_t and location of *kink* type of singularities on Ω_t by an algorithm (ii) which takes small computing time. Both are achieved by KCL-based shock ray theory. We note that the KCL method reduces the number of independent variables by one, which reduces the computing time drastically. A conservative formulation in KCL also has the advantage of using modern shock-capturing algorithms and tracking shock surfaces in the ray coordinates $(\boldsymbol{\xi}, t)$. When we map the results onto the $\boldsymbol{x}$-space, we get a very sharp location of Ω_t and the kink lines on it. Thus, both Ω_t and kinks on it are automatically tracked. We support the theory in the book by referring to extensive numerical results of our group. These results, some of which have been reproduced here, have been published in important journals during the last 16 years.

Summation convention: Throughout the book, we shall use the summation convention that a repeated symbol in subscripts and superscripts in a term will mean summation over the range of the symbol. The range of the symbols $\alpha, \beta,$ *and* γ will be $1, 2, \ldots, d$ and those of $i, j,$ *and* k will be $1, 2, \ldots, n$. We shall also need subscripts p and q with range $1, 2, \ldots, d-1$.

List of symbols	
d	dimension of the physical space.
$\boldsymbol{x}$	$(x_1, x_2, \ldots, x_d) \in \mathbb{R}^d$.
$(\boldsymbol{x}, t) = (x_1, x_2, t)$	for 2-D KCL.
$(\boldsymbol{x}, t) = (x_1, x_2, x_3, t)$	for 3-D KCL.
$(\boldsymbol{x}, t) = (x_1, x_2, \ldots x_d, t)$	for d-D KCL.
Ω: $\varphi(\boldsymbol{x}, t) = 0$	a surface in $(\boldsymbol{x}, t)$ space.
Ω_t: $\varphi(\boldsymbol{x}, t) = 0, t =$ constant	a moving surface in $\boldsymbol{x}$-space at a fixed time t.
$\mathit{\Omega}$	mean curvature of Ω_t.
$\boldsymbol{n}$	unit normal of Ω_t.
θ	the angle $\boldsymbol{n}$ makes with x-axis for 2-D KCL.
$\boldsymbol{\chi}$	ray velocity, $= m\boldsymbol{n}$ for isotropic evolution of Ω_t.
a	characteristic velocity of the wave equation and also sound velocity in a gas.
a_*	common critical speed in a gas on two sides of a shock in a gas, $= \frac{\gamma-1}{\gamma+1}\hat{q}^2$.
$\hat{q}$	common limit speed on two sides of a shock in a gas.
A	shock speed in gas dynamics relative to the flow velocity ahead of the shock.
$\boldsymbol{q}, \rho, p, \sigma$	velocity, density, pressure, entropy of a gaseous medium.
$c_1, c_2, \ldots, c_n$	eigenvalues of a hyperbolic system.
KCL	kinematical conservation laws.
PDE	partial differential equation.
WNLRT	weakly nonlinear Ray theory.
w_i	characteristic variable of ith characteristic field.
$\tilde{w}, w$	dimensional and non-dimensional amplitude of a nonlinear wavefront.
m	normal velocity of Ω_t and is the metric associated with t in ray coordinates.
M	normal velocity or Mach number of shock front and is the metric associated with t in ray coordinates.
$\boldsymbol{\pi} = (\pi_1, \pi_2, \ldots, \pi_{n-1})$	Riemann invariants.
(ξ, t)	ray coordinates for 2-D KCL.
g	metric associated with ξ.
$(\xi_1, \xi_2, \ldots, \xi_{d-1}, t)$	ray coordinates for d-D KCL.

List of symbols	
g_i	metric associated with $\xi_i, i = 1, 2, \ldots, (d-1)$.
$\boldsymbol{u}, \boldsymbol{v}$	unit tangent vectors on Ω_t in direction of the coordinates ξ_1 and ξ_2 for 3-D KCL.
$\boldsymbol{u}_1, \boldsymbol{u}_2, \ldots \boldsymbol{u}_{d-1}$	unit tangent vectors on Ω_t in direction of the coordinates $\xi_1, \xi_2 \ldots, \xi_{d-1}$ for d-D KCL.
$\boldsymbol{L}$	$\nabla - \boldsymbol{n}\langle \boldsymbol{n}, \nabla\rangle$.
χ	$\cos\chi = \langle \boldsymbol{u}, \boldsymbol{v}\rangle,\ 0 < \chi < \pi$ for WNLRT.
Ψ	$\cos\Psi = \langle \boldsymbol{U}, \boldsymbol{V}\rangle,\ 0 < \chi < \pi$ for SRT.
$\mathcal{S}$	shock surface in $(\boldsymbol{x}, t)$-space or $(\boldsymbol{\xi}, t)$-space represented by $\mathcal{S}(\boldsymbol{x}, t) = 0$ or $\mathcal{S}(\boldsymbol{\xi}, t) = 0$.
$\mathcal{S}_t$	shock front in $\boldsymbol{x}$-space or $\boldsymbol{\xi}$-space.
S	velocity of propagation of a shock front $\mathcal{S}_t$.
$\mathcal{K}$	kink surface on Ω in $(\boldsymbol{x}, t)$-space.
$\mathcal{K}_t$	kink surface, called simply as kink, on Ω_t in $\boldsymbol{x}$-space.
$\boldsymbol{E} = (E_1, E_2, \ldots, E_{d-1})$	unit normal to the shock front $\mathcal{S}_t$ in $\boldsymbol{\xi}$-space.
$[f] := (f+) - (f-)$	jump of a quantity across a shock surface $\mathcal{S}$ in ($\boldsymbol{x},\boldsymbol{t}$)-space or in $(\boldsymbol{\xi}, t)$-space.
$a_{ij}, b_{ij}^{(1)}, b_{ij}^{(2)}$	components of 7×7 matrices $A, B^{(1)}$ and $B^{(2)}$ in PDE of 3-D WNLRT (10.23).
$\tilde{a}_{ij}, \tilde{b}_{ij}^{(1)}, \tilde{b}_{ij}^{(2)}$	components of 8×8 matrices $\tilde{A}, \tilde{B}^{(1)}$ and $\tilde{B}^{(2)}$ in PDE of 3-D SRT, (10.28).
$\tilde{w}$	amplitude of a nonlinear wavefront in a polytropic gas (see relation (5.10)).
$\mathcal{A}$	ray tube area.
$\frac{\partial}{\partial \eta_\beta^\alpha}$	$n_\beta \frac{\partial}{\partial x_\alpha} - n_\alpha \frac{\partial}{\partial x_\beta}$.
$\lambda_1, \lambda_2(= -\lambda_1), \lambda_3 = \ldots = \lambda_7 = 0$	eigenvalues of 3-WNLRT (10.23).
$\lambda_1', \lambda_2'(= -\lambda_1'), \lambda_3' = \ldots = \lambda_7' = 0$	eigenvalues of the frozen system with $(\boldsymbol{u}, \boldsymbol{v}) = (\boldsymbol{u}', \boldsymbol{v}')$, where $\langle \boldsymbol{u}', \boldsymbol{v}'\rangle = 0$, see (10.28).
(η_1, η_2)	orthogonal coordinates on Ω_t with unit tangent vectors $\boldsymbol{u}', \boldsymbol{v}'$ frozen at a point P_0.
(g_1', g_2')	metrics associated with η_1 and η_2 frozen at a point P_0.
$\gamma_1, \gamma_2, \delta_1, \delta_2, e_1, e_2, e_1', e_2'$	coefficients occurring in Sect. 10.1.

Chapter 2
Scalar First-Order PDE

The most general form of a first-order PDE is (1.1). We shall consider a particular case of this in the next section.

2.1 Quasilinear PDE

A first-order quasilinear PDE for $u : \mathbb{R}^{d+1} \to \mathbb{R}$ is

$$a(\boldsymbol{x}, t, u)u_t + \langle \boldsymbol{b}(\boldsymbol{x}, t, u), \nabla_x u \rangle = c(\boldsymbol{x}, t, u), \tag{2.1}$$

where $a : \mathbb{R}^{d+2} \to \mathbb{R}$, $\boldsymbol{b} : \mathbb{R}^{d+2} \to \mathbb{R}^d$, $c : \mathbb{R}^{d+2} \to \mathbb{R}$ and we assume that

$$a,\ \boldsymbol{b},\ c\ are\ C^1\ functions\ on\ a\ domain\ in\ \mathbb{R}^{d+2}. \tag{2.2}$$

When at least one of $|\nabla_u a|$ and $|\nabla_u b_\alpha|$ is non-zero, the Eq. (1.1) is **quasilinear**.

Take a solution[1] $u(\boldsymbol{x}, t)$ of (2.1), then the functions $a(\boldsymbol{x}, t, u(\boldsymbol{x}, t))$ and $\boldsymbol{b}(\boldsymbol{x}, t, u(\boldsymbol{x}, t))$ can be treated simply as functions of $\boldsymbol{x}$ and t.

Consider a curve $\gamma : (\boldsymbol{x}(\sigma), t(\sigma))$ in $\mathbb{R}^{d+1}$ with tangent direction given by the vector $(\boldsymbol{b}, a)$, then with a proper choice of σ

$$\frac{d\boldsymbol{x}}{d\sigma} = \boldsymbol{b}(\boldsymbol{x}, t, u(\boldsymbol{x}, t)), \quad \frac{dt}{d\sigma} = a(\boldsymbol{x}, t, u(\boldsymbol{x}, t)). \tag{2.3}$$

Along such a curve γ

$$\frac{du}{d\sigma} = \frac{dt}{d\sigma}u_t + \langle \frac{d\boldsymbol{x}}{d\sigma}, \nabla_x u \rangle = a(\boldsymbol{x}, t, u(\boldsymbol{x}, t))u_t + \langle \boldsymbol{b}(\boldsymbol{x}, t, u(\boldsymbol{x}, t)), \nabla_x u \rangle.$$

[1]For a first-order linear PDE, we can follow the same steps as given in this section except that we do not have to take a known solution.

P. Prasad, *Propagation of Multidimensional Nonlinear Waves and Kinematical Conservation Laws*, Infosys Science Foundation Series,
https://doi.org/10.1007/978-981-10-7581-0_2

Using (2.1), we get

$$\frac{du}{d\sigma} = c(\boldsymbol{x}, t, u(\boldsymbol{x}, t)). \tag{2.4}$$

Equations (2.3) and (2.4) are true for every solution $u(\boldsymbol{x}, t)$. Hence, it follows that for any solution $u(\boldsymbol{x}, t)$ of (2.1), there are curves in $(\boldsymbol{x}, t)$-space which are given by the

$$\frac{d\boldsymbol{x}}{d\sigma} = \boldsymbol{b}(\boldsymbol{x}, t, u), \quad \frac{dt}{d\sigma} = a(\boldsymbol{x}, t, u) \tag{2.5}$$

and along these curves

$$\frac{du}{d\sigma} = c(\boldsymbol{x}, t, u). \tag{2.6}$$

For a quasilinear equation, (2.5) and (2.6) form a coupled system of $d+2$ equations.

Definition 2.1.1 We call the $d + 1$ equations (2.5), **characteristic equations** of (2.1).

Definition 2.1.2 We call Eq. (2.6), **compatibility condition** of (2.1).

Definition 2.1.3 We call an integral curve of Eqs. (2.5) and (2.6) in $(d + 2)$-dimensional $(\boldsymbol{x}, t, u)$-space, **Monge curve** of (2.1).

Definition 2.1.4 We call the projection of a Monge curve on $(d + 1)$-dimensional $(\boldsymbol{x}, t)$-space, **characteristic curve** of (2.1).

Remark 2.1.5 Monge curves (and Monge strips to be defined for nonlinear equations in the next section) in $(\boldsymbol{x}, t, u)$-space of independent variables $\boldsymbol{x}, t$ and dependent variable u have been called characteristic curves (and characteristic strips) by all other authors but we reserve the word 'characteristics' to be associated with the projections of Monge curves on the space of independent variables consistent with the use of this word for a higher order equation or a systems of equations, see also [47].

For a **linear or semilinear** PDE, the coefficient a and $\boldsymbol{b}$ in (2.1) are independent of u and hence the characteristic Eqs. (2.5) decouple from the compatibility condition (2.6). In this case, the characteristic curves form a d parameter family of curves in $(\boldsymbol{x}, t)$-space. For a quasilinear equation, Eqs. (2.5) and (2.6) are coupled and hence characteristic curves form a $(d + 1)$-parameter family of curves in $(\boldsymbol{x}, t)$-space.

Solution of a Cauchy Problem: Let us write the Cauchy data (1.5) for the Eq. (1.1) in a parametric form

$$\boldsymbol{x} = \boldsymbol{\eta} := (\eta_1, \eta_2, \cdots, \eta_d), \quad t = 0, \quad u(\boldsymbol{x}, 0) = u_0(\boldsymbol{\eta}). \tag{2.7}$$

Since the coefficients a and $\boldsymbol{b}$ of (2.1) are C^1 functions of their arguments, solution of (2.5) and (2.6) exists for each $\boldsymbol{\eta}$ locally in a neighbourhood of $\sigma = 0$. We write this solution in the form

$$x = X(\sigma, \eta), \quad t = T(\sigma, \eta) \tag{2.8}$$

and

$$u = U(\sigma, \eta). \tag{2.9}$$

It also follows that $\frac{\partial X_\alpha}{\partial \sigma}|_{\sigma=0} = b_\alpha$, $\frac{\partial t}{\partial \sigma}|_{\sigma=0} = a$ and $\frac{\partial X_\alpha}{\partial \eta_\beta}|_{\sigma=0} = \delta_{\alpha\beta}$, $\frac{\partial t}{\partial \eta_\alpha}|_{\sigma=0} = 0$, for $\alpha, \beta = 1, 2, \cdots, d$; where $\delta_{\alpha\beta}$ are Kronecker deltas. So the Jacobian

$$J|_{\sigma=0} := \frac{\partial(X_1, X_2, \cdots, X_d, t)}{\partial(\eta_1, \eta_2, \cdots, \eta_d, \sigma)}|_{\sigma=0} = a. \tag{2.10}$$

When $a \neq 0$, the Jacobian J of transformation from (σ, η)-space to (x, t)-space is non-zero at $\sigma = 0$ and we can solve σ and η from (2.8) as functions of x and t in a neighbourhood of a point on $t = 0$ in the form

$$\sigma = \sigma(x, t), \quad \eta = \eta(x, t) \tag{2.11}$$

and substitute in (2.9) to get a function

$$u(x, t) := U(\sigma(x, t), \eta(x, t)). \tag{2.12}$$

We can prove that this function is the unique solution of the Cauchy problem.

Now we proceed to define **Cauchy problem** with Cauchy data u_0 prescribed on an arbitrary surface γ in (x, t)-space.

Definition 2.1.6 Let a parametric representation of γ be

$$\gamma : x = x_0(\eta), \quad t = t_0(\eta), \quad \eta \ \ in\ a\ domain\ \subset \mathbb{R}^d \tag{2.13}$$

on which the Cauchy data is prescribed as

$$u_0 = u_0(\eta).$$

The Cauchy problem for a first-order PDE is to find a solution $u(x, t)$ in a neighbourhood of γ such that $u(x_0(\eta), t_0(\eta)) = u_0(\eta)$.

Definition 2.1.7 A d-dimensional surface $\Omega : \varphi(x, t) = 0$ in (x, t)-space is a **characteristic surface** of a PDE (or a system of PDEs (1.4)) if characteristic curves[2] at every point of it are tangential to it.

A characteristic surface Ω, a d-dimensional surface, is generated by a motion of a family of $d - 1$ parameter family of characteristic curves.

The condition $a \neq 0$ also implies that the characteristic curves given by (2.5) and (2.6) are transversal to the plane $t = 0$. Hence the plane, where the Cauchy data is prescribed in the problem (2.7), is a non-characteristic plane.

[2]Defined in Sect. (4.2.1) for a system of equations.

Remark 2.1.8 Consider a Cauchy problem in which the data is prescribed not on $t = 0$ (i.e. not as in (1.5)) but on any smooth manifold $\gamma : \boldsymbol{x} = \boldsymbol{x}_0(\eta),\ t = t_0(\eta)$ in $(\boldsymbol{x}, t)$-space as in (2.13). Let $a, \boldsymbol{b}$ *and* c be C^1 functions. If γ is not a characteristic surface, we can prove that the solution of this non-characteristic Cauchy problem exists and is unique.

Definition 2.1.7 of a characteristic surface $\Omega : \varphi(\boldsymbol{x}, t) = 0$, using (2.5), implies that

$$a\varphi_t + \langle \boldsymbol{b}, \nabla_x \varphi \rangle = 0 \quad on\ \Omega \tag{2.14}$$

which, due to (2.5), implies

$$\frac{d\varphi}{dt} = 0. \tag{2.15}$$

In the derivation of the (2.14), we have used characteristic equations for a known solution $u(\boldsymbol{u}, t)$, otherwise these curves are not defined except in the case of a semi-linear equation. But this solution must satisfy the compatibility condition (2.6) along each one of the characteristic curves on Ω. Hence, if a choose a Cauchy problem in which the data is prescribed on a characteristic surface, the data cannot be prescribed arbitrarily on Ω. Thus, a characteristic Cauchy problem may not have a solution but if it has a solution it can be shown that it will have infinity of solutions.

For a quasilinear equation, the characteristic PDE (2.14) depends on u, since a and $\boldsymbol{b}$ depend on u.

Genuine nonlinearity is an important concept, which can be explained in a very simple way for a scalar first-order quasilinear PDE $u_t + f(x, t, u)u_x = g(x, t, u)$ in two independent variables, where f and g are known smooth functions. The slope $\frac{dx}{dt} = f(x, t, u)$ of a characteristic curve in (t, x)-plane represents the **velocity** of propagation with which a value u **propagates** along the characteristic according to the compatibility condition $\frac{du}{dt} = g(x, t, u)$.

Definition 2.1.9 We define the **characteristic field** to be **genuinely nonlinear characteristic field** if $f^{'}(u) \neq 0$, for all u.

Thus, in a characteristic field which is not genuinely nonlinear, the velocity of propagation of a value u along a characteristic curve is independent of the value of u. We shall define genuine nonlinearity for a system of equations in Chap. 4.

General Solution of PDE (2.1): We give here a few steps to find a general solution (for detailed theory see [13, 47]).

1. Each of the first integrals of the characteristic equations and compatibility condition gives an **integral surface** of the (2.1). Here, integral surface means the surface represented by a solution $u = f(x, t)$ in $(\boldsymbol{x}, t, u)$-space.
2. Let a complete set of $d + 1$ independent first integrals of (2.25) and (2.26) be $f_\alpha(\boldsymbol{x}, t, u) = \text{constant}, c_\alpha$, say; $\alpha = 1, 2, \cdots, d + 1$.

3. A **general solution** of (2.1) is

$$\varphi\{f_1(\boldsymbol{x}, t, u), f_2(\boldsymbol{x}, t, u), \cdots, f_{d+1}(\boldsymbol{x}, t, u)\} = 0,$$

where φ is the arbitrary function of $d+1$ arguments.

A general solution is really not a solution but if we are able to solve it for u in terms of $\boldsymbol{x}$ and t, we get a solution of the PDE containing an arbitrary function. It is possible to show that we can use a general solution to solve a Cauchy problem.

2.2 Examples and Some General Results

Let us explain all concepts developed so far with the help of some examples.

Example 2.2.1 Consider a PDE

$$u_t + (\alpha_1 + \alpha_2 u)u_{x_1} + (\beta_1 + \beta_2 u)u_{x_2} = mu, \tag{2.16}$$

where α_1, α_2, β_1, β_2 and m are constant.

Since the coefficient of u_t is 1, we can replace σ by t in (2.5) and (2.6). The characteristic equations and compatibility condition are

$$\frac{dx_1}{dt} = \alpha_1 + \alpha_2 u, \quad \frac{dx_2}{dt} = \beta_1 + \beta_2 u, \tag{2.17}$$

and

$$\frac{du}{dt} = mu. \tag{2.18}$$

General solution (which consists of 3 first integrals) of ordinary differential equations (2.17) and (2.18) contains three arbitrary constants x_{10}, x_{20} and u_0, and is given by

$$x_1 = x_{10} + \alpha_1 t + \frac{\alpha_2 u_0}{m}(e^{mt} - 1), \quad x_2 = x_{20} + \beta_1 t + \frac{\beta_2 u_0}{m}(e^{mt} - 1), \tag{2.19}$$

$$u = u_0 e^{mt} \tag{2.20}$$

This gives three-parameter family of **Monge curves** in $(\boldsymbol{x}, t, u)$-space. The equations in (2.19) give the three-parameter family of **characteristic curves** in $(\boldsymbol{x}, t)$-space.

We write the solution (2.19) and (2.20) with the arbitrary constants on the right-hand side

$$x_1 - \alpha_1 t - \frac{\alpha_2 u}{me^{mt}}(e^{mt} - 1) = x_{10}, \quad x_2 - \beta_1 t - \frac{\beta_2 u}{me^{mt}}(e^{mt} - 1) = x_{20}, \quad ue^{-mt} = u_0. \tag{2.21}$$

A general solution of (2.16) is

$$\varphi\{ue^{-mt}, \; x_1 - \alpha_1 t - \frac{\alpha_2 u}{me^{mt}}(e^{mt} - 1), \; x_2 - \beta_1 t - \frac{\beta_2 u}{me^{mt}}(e^{mt} - 1)\} = 0 \tag{2.22}$$

where φ is the arbitrary function of three arguments.

2-D Burgers' Equation This is a particular case of Eq. (2.16) when $m = 0$:

$$u_t + (\alpha_1 + \alpha_2 u)u_{x_1} + (\beta_1 + \beta_2 u)u_{x_2} = 0. \tag{2.23}$$

All corresponding results of this equation can be obtained either directly or as particular cases of the results of (2.16) by taking limit as $m \to 0$. We note that $\lim_{m\to 0} \frac{1}{m}(e^{mt} - 1) = t$. A general solution of (2.23) is

$$\varphi(u, \; x_1 - \alpha_1 t - \alpha_2 ut, \; x_2 - \beta_1 t - \beta_2 ut) = 0 \tag{2.24}$$

where φ is an arbitrary function or in an another form

$$u = u_0(x_1 - \alpha_1 t - \alpha_2 ut, \; x_2 - \beta_1 t - \beta_2 ut), \tag{2.25}$$

where u_0 is an arbitrary function of two arguments.

Example 2.2.2 For a PDE

$$u_t + u_x = 0 \tag{2.26}$$

we can derive from the characteristic equations and compatibility condition, the following results including a general solution as

$$x - ct = c_1, \quad u = c_2, \; and \; u = f(x - ct), \tag{2.27}$$

where c_1 and c_2 are constant and f is an arbitrary C^1 function.

Consider a Cauchy problem in which data is prescribed on a curve $\gamma : x = x_0(t)$, where $x_0 \in C^1$. A parametric representation of the Cauchy data is

$$x = x_0(\eta), \quad t = \eta, \quad u = u_0(\eta) \tag{2.28}$$

and the solution of the characteristic equations and compatibility condition satisfying (2.28) is of the form

$$x - ct = x_0(\eta) - c\eta, \quad u = u_0(\eta). \tag{2.29}$$

An example of a **Non-Characteristic Cauchy Problem:** We can solve η uniquely in terms of x and t from the first equation in (2.29) if $x_0'(\eta) \neq c$. The solution is $\eta = g(x - ct)$, $g \in C^1$. This can be substituted in the second equation to get

$$u(x, y) = u_0(\eta) = u_0(g(x - ct)), \tag{2.30}$$

which is a C^1 function satisfying (2.26). Thus, the solution of the Cauchy problem (2.29) for (2.26) exists and is unique if $x_0'(\eta) \neq c$, i.e. the datum curve is nowhere tangential to a characteristic curve.

An example of a **Characteristic Cauchy Problem**: Let the datum curve be a characteristic curve $x = ct + 1$. The data $u_0(t)$ prescribed on this line must be a constant, say $u_0(t) = a$. Now, we can verify that the solution is given by

$$u(x, y) = a + (x - ct - 1)h(x - ct), \tag{2.31}$$

where $h(\eta)$ is an *arbitrary* C^1 function of just one argument. This verifies a general property that the solution of a characteristic Cauchy problem, when it exists, it is **not unique**.

Example 2.2.3 Consider a Cauchy initial value problem for the Burgers' equation

$$u_t + uu_x = 0 \tag{2.32}$$

with initial data (1.5). This is a non-characteristic Cauchy problem. The solution of this problem is given by

$$u(x, y) = u_0(x - ut). \tag{2.33}$$

Solving (2.33) for u is a difficult problem but when $u_0 \in C^1$, the solution exists and is unique in a neighbourhood of the initial line $t = 0$. From (2.33) it is clear that if u_0 is bounded, u is also bounded as long as the solution exists. Let us try to find out the derivative u_x of the solution in terms of the initial data. Differentiating the PDE (2.32) with respect to x, we write the characteristic compatibility conditions for u and u_x as

$$\frac{du}{dt} = 0 \quad and \quad \frac{du_x}{dt} = -(u_x)^2 \quad along \quad \frac{dx}{dt} = u. \tag{2.34}$$

Solution for u_x with initial data $u_x(x, 0) = u_0'(x)$ at $t = 0$, where $'$ represents derivative, is

$$u_x = \frac{u_0'(\xi)}{1 + tu_0'(\xi)}, \quad \xi = x - ut. \tag{2.35}$$

If the initial data is such that $u_0'(x) < 0$ on some interval of the x-axis, there exists a time $t_c > 0$ such that as $t \to t_c-$, the derivative $u_x(x, t)$ of the solution tends to $-\infty$ for some value of x and thus the genuine solution cannot be continued beyond at $t = t_c$. The critical time t_c is given by

$$t_c = -\frac{1}{\min_{\xi \in \mathbb{R}} u_0'(\xi)} > 0. \tag{2.36}$$

If $u_0'(x) > 0$ for all $x \in \mathbb{R}$, $t_c < 0$ and the relation (2.33) gives a genuine solution of the initial value problem for all $t > 0$.

This example highlights a very important aspect of solutions of quasilinear PDEs. Even if the initial data $u_0(x)$ is infinitely differentiable, the **genuine nonlinearity** (see definition (2.1.9)) present in the (2.32), causes a blow up not of u but of the derivative u_x. For details see [13, 17, 47], especially [42]. Discussion of the solution after t_c is important both from the point of view of mathematics and physics. This will be taken up in the next chapter dealing with the theory of single conservation law.

2.3 Nonlinear Equation

The most general form of first-order equation is (1.1), where F is a nonlinear function of $\nabla_{\boldsymbol{x}} u$ and u_t. We introduce symbols

$$p_\alpha = u_{x_\alpha}, \quad \boldsymbol{p} = \nabla_{\boldsymbol{x}} u, \quad q = u_t, \tag{2.37}$$

and we write (1.1) in the form $F(\boldsymbol{x}, t, u, \boldsymbol{p}, q) = 0$. We assume that $F \in C^2$ in a domain of $(2d + 3)$-dimensional space. We take a known C^2 solution $u(\boldsymbol{x}, t)$ and differentiate (1.1) with respect to x_α. This gives

$$F_{x_\alpha} + u_{x_\alpha} F_u + (p_\beta)_{x_\alpha} F_{p_\beta} + q_{x_\alpha} F_q = 0. \tag{2.38}$$

Using $(p_\beta)_{x_\alpha} = (p_\alpha)_{x_\beta}$, $q_{x_\alpha} = (p_{x_\alpha})_t$ and rearranging the terms, we get

$$(p_\alpha)_{x_\beta} F_{p_\beta} + (p_{x_\alpha})_t F_q = -F_{x_\alpha} - p_\alpha F_u \tag{2.39}$$

in which the quantities F_{p_β}, F_q are now known functions of x and t. This is a beautiful result, the x_α-derivative of u, namely p_α is differentiated in the direction $(F_{p_1}, F_{p_2}, \cdots, F_{p_d}, F_q)$ in $(\boldsymbol{x}, t)$-space. For the known solution $u(\boldsymbol{x}, t)$, consider a d parameter family of curves in $(\boldsymbol{x}, t)$-space given by

$$\frac{dx_\alpha}{d\sigma} = F_{p_\alpha}, \quad \frac{dt}{d\sigma} = F_q. \tag{2.40}$$

Along these curves, (2.39) gives

$$\frac{dp_\alpha}{d\sigma} = -F_{x_\alpha} - p_\alpha F_u. \tag{2.41}$$

Similarly, differentiating (1.1) with respect to t and following the same procedure, we find that along the curves given by (2.40), we have

$$\frac{dq}{d\sigma} = -F_t - qF_u. \tag{2.42}$$

The rate of change of u along these curves is

$$\frac{du}{d\sigma} = u_{x_\alpha}\frac{dx_\alpha}{d\sigma} + u_t\frac{dt}{d\sigma} = p_\alpha F_{p\alpha} + qF_q. \tag{2.43}$$

Note that (2.40)–(2.43) form a complete system of $2d + 3$ ODEs for $2d + 3$ quantities $x_1, x_2, \cdots, x_d, t, u, p_1, p_2, \cdots, p_d$ and q for any solution $u(\boldsymbol{x}, t)$ we take for their derivation. Thus, we find a beautiful set of equations, called **Charpit's equations** consisting of $d+1$ *characteristic equations* (2.40) and $d+2$ compatibility conditions (2.41)–(2.43). Given a set of values $(u_0, \boldsymbol{p}_0, q_0)$ at any point $(\boldsymbol{x}_0, t_0)$, so that $(\boldsymbol{x}_0, t_0, u_0, \boldsymbol{p}_0, q_0)$ belongs to the domain in which F is defined, we can find a unique local solution of Charpit's equations. Since the system is autonomous, the set of solutions of the Charpit's equations form a $2d + 2$ parameter family of curves in $(\boldsymbol{x}, t)$-space.

Remark 2.3.1 Every solution $(\boldsymbol{x}(\sigma), t(\sigma), u(\sigma), \boldsymbol{p}(\sigma), q(\sigma))$ of the Charpit's equations satisfies the **strip condition**

$$\frac{du}{d\sigma} = p_\alpha(\sigma)\frac{dx_\alpha}{d\sigma} + q(\sigma)\frac{dt}{d\sigma} \tag{2.44}$$

on the curve $(\boldsymbol{x}(\sigma), t(\sigma), u(\sigma))$ in $(\boldsymbol{x}, t, u)$-space. For a geometrical interpretation of the strip condition, see [13, 47].

There is an interesting result, which says that *the function F is constant on any integral curve of Charpit's equations in $(\boldsymbol{x}, t, u, \boldsymbol{p}, q)$-space*. The proof is very simple. This means that though not every solution of the Charpit's equations satisfies $F(\boldsymbol{x}(\sigma), t(\sigma), u(\sigma), \boldsymbol{p}(\sigma), q(\sigma)) = 0$, if we choose $u_0, \boldsymbol{p}_0$ and q_0 at $(\boldsymbol{x}_0, t_0)$ such that $F(\boldsymbol{x}_0, t_0, u_0, \boldsymbol{p}_0, q_0) = 0$, then $F = 0$ for all values of σ.

Definition 2.3.2 A set of $2d + 3$ ordered functions $(\boldsymbol{x}(\sigma), t(\sigma), u(\sigma), \boldsymbol{p}(\sigma), q(\sigma))$ satisfying the Charpit equations (2.40)–(2.43) and $F(\boldsymbol{x}(\sigma), t(\sigma), u(\sigma), \boldsymbol{p}(\sigma), q(\sigma)) = 0$ is called a **Monge strip** of the PDE (1.1).

The condition $F(\boldsymbol{x}(\sigma), t(\sigma), u(\sigma), \boldsymbol{p}(\sigma), q(\sigma)) = 0$ imposes a relation between the $2d+2$ parameters of the set of all solutions of the Charpit's equations. Therefore, the of Monge strips form a $2d + 1$ parameter family of strips in $(\boldsymbol{x}, t, u)$-space.

Definition 2.3.3 Given a Monge strip $(\boldsymbol{x}(\sigma), t(\sigma), u(\sigma), \boldsymbol{p}(\sigma), q(\sigma))$, the base curve in $(\boldsymbol{x}, t)$-space given by $(\boldsymbol{x}(\sigma), t(\sigma))$ is called a **characteristic curve** of the PDE (1.1).

Remark 2.3.4 Characteristic curves of a linear first-order PDE form a d parameter family of curves in $(\boldsymbol{x}, t)$-plane. Those of a quasilinear equation form a $d+1$ parameter family of curves. Finally, those of a nonlinear equation (1.1) form a $2d+1$ parameter family of curves.

Solution of a Cauchy Problem:
We state an algorithm to solve a Cauchy problem. The proof, which guarantees that this algorithm indeed gives a unique solution of the Cauchy problem, can be found in [13, 17, 47].

For the nonlinear PDE (1.1), we need to solve the Charpit's equations (2.40)–(2.43) and for this, we need initial values $\boldsymbol{p}_0(\eta)$ and $q_0(\eta)$ on γ in addition to the values $\boldsymbol{x}_0(\eta)$, $t_0(\eta)$ and $u_0(\eta)$ given in the Cauchy data in Definition 2.1.6. First, we note that these initial values must satisfy the PDE, i.e.

$$F(\boldsymbol{x}_0(\eta), t_0(\eta), u_0(\eta), \boldsymbol{p}_0(\eta), q_0(\eta)) = 0. \tag{2.45}$$

Further, differentiating $u(\boldsymbol{x}_0(\eta), t_0(\eta)) = u_0(\eta)$ with respect to η_α we get d more relations

$$p_{\beta 0}\frac{\partial x_{\beta 0}}{\partial \eta_\alpha} + q_0\frac{\partial t_0}{\partial \eta_\alpha} = \frac{\partial u_0}{\partial \eta_\alpha}. \tag{2.46}$$

We take $d+1$ functions $\boldsymbol{p}_0(\eta)$ and $q_0(\eta)$ satisfying (2.45)–(2.46) and complete the initial data for (2.40)–(2.43) at $\sigma = 0$ as

$$\boldsymbol{x}(0,\eta) = \boldsymbol{x}_0(\eta),\ t(0,\eta) = t_0(\eta),\ u(0,\eta) = u_0(\eta),\ \boldsymbol{p}(0,\eta) = \boldsymbol{p}_0(\eta),\ q(0,\eta) = q_0(\eta). \tag{2.47}$$

Now, we solve the Charpit's equations (2.40)–(2.43) with initial data (2.47) and obtain

$$\boldsymbol{x} = \boldsymbol{X}(\sigma,\eta),\ t = T(\sigma,\eta),\ u = U(\sigma,\eta),\ \boldsymbol{p} = \boldsymbol{P}(\sigma,\eta),\ q = Q(\sigma,\eta). \tag{2.48}$$

From the first two relations in (2.48), we solve σ and η as functions of $\boldsymbol{x}$ and t and substitute in the third relation to get the solution $u(\boldsymbol{x}, t)$ of the Cauchy problem. The following theorem (stated without proof) assures that the above algorithm indeed gives a local solution of the Cauchy problem.

Theorem 2.3.5 *Consider a Cauchy problem for the PDE (1.1) with Cauchy data $u_0(\eta)$ prescribed on a surface γ as given in definition (2.1.6). Let*
(i) the $F(\boldsymbol{x}, t, u, \boldsymbol{p}, q) \in C^2(D_3)$, where D_3 is a domain in $(\boldsymbol{x}, t, u, \boldsymbol{p}, q)$-space,
(ii) the functions $\boldsymbol{x}_0(\eta), t_0(\eta), u_0(\eta) \in C^2$,
(iii) $\boldsymbol{p}_0(\eta)$ and $q_0(\eta)$ be $d+1$ functions satisfying Eqs. (2.45) and (2.46) such that they are C^1 and the set $\{\boldsymbol{x}_0(\eta), t_0(\eta), u_0(\eta), \boldsymbol{p}_0(\eta), q_0(\eta)\} \in D_3$ and
(iv) the transversality condition that the vector $(\nabla_{\boldsymbol{p}} F, F_q)$ is nowhere tangential to γ is satisfied.

Then, we can find a domain D in $(\boldsymbol{x}, t)$-space containing the datum surface γ and a unique solution of the Cauchy problem in D.

Characteristic Cauchy Problem:
We remind the general definition of a characteristic surface in Definition 2.1.7, which remains valid also for a nonlinear PDE. Important point, for the existence and uniqueness of the Cauchy problem in the above theorem, is that due to the transversality condition in (iv), the datum surface γ is nowhere tangential to a characteristic curve. If γ is a characteristic surface and the data $u_0(\eta)$ is restricted according to the compatibility conditions (2.41)–(2.43), the solution of the Cauchy problem is non-unique, in fact infinity of solutions exist.

2.4 Example: Eikonal Equation

Eikonal equation plays an important role in the theory of wave propagation in particular and a propagating surface in general. We discuss here its simplest case.

Example 2.4.1 Consider a pulse generated by a one-parameter family of wavefronts moving into a uniform 2-D medium with a constant normal velocity a_0 (see Sect. 3.5 for an example of one-parameter family of wavefronts). Let the successive positions of the one parameter family of wavefronts Ω_t^k be represented by an equation $u(x_1, x_2, t, k) = 0$, where and k is the parameter. Since the velocity of propagation a_0 is given by $a_0 = -\frac{u_t}{({u_{x_1}}^2 + {u_{x_2}}^2)^{1/2}}$, the function $u(x_1, x_2, t, k)$ satisfies an eikonal equation[3]

$$F := a_0\sqrt{p_1^2 + p_2^2} + q = 0; \quad p_1 = u_{x_1}, \ p_2 = u_{x_2}, \ q = u_t. \tag{2.49}$$

Let us take a one-parameter family curves Ω_0^k, representing initial positions of the family of wavefronts, given by

$$\Omega_0^k : \quad u_0(x_1, x_2) \equiv \varphi_0(x, y) - k = 0, \quad k = constant. \tag{2.50}$$

We shall find the successive positions of the wavefronts Ω_t^k, originating from each of these curves Ω_0^k, as they evolve. Consider the following two cases of the initial data $u(\boldsymbol{x}, 0) = u_0(\boldsymbol{x})$ for the PDE (2.49).
Case 1 Let $\varphi_0(x, y) = \alpha x_1 + \beta x_2$, i.e. we take

$$u_0 = \alpha x_1 + \beta x_2 - k, \quad \alpha, \beta = constants, \quad \alpha^2 + \beta^2 = 1. \tag{2.51}$$

We choose a parametric representation of the Cauchy data at $\sigma = 0$ as

[3]This eikonal equation is also obtained by factorizing the left-hand side of the characteristic PDE (4.32) of the wave equation (4.31) and equating it to zero.

$$x_{10} = \eta_1, \quad x_{20} = \eta_2, \quad t_0 = 0, \quad u_0 = \alpha\eta_1 + \beta\eta_2 - k \tag{2.52}$$

and deduce the following initial values of p_{10} *and* p_{20} from (2.45) and (2.46) with F given by (2.49):

$$p_{10} = \alpha, \quad p_{20} = \beta, \quad q_0 = -a_0. \tag{2.53}$$

Since the coefficient of q in (2.49) is 1, we can replace σ by t in the characteristic equations and compatibility conditions. Solution of these with the values in (2.52) and (2.53) at $t = 0$ is

$$x_1 = \eta_1 + \alpha a_0 t, \ x_2 = \eta_2 + \beta a_0 t, \ p_1 = \alpha, \ p_2 = \beta, \ q_0 = -a_0, \ u = \alpha\eta_1 + \beta\eta_2 - k. \tag{2.54}$$

Elimination of η_1 and η_2 now gives the solution u as

$$u = \alpha x_1 + \beta x_2 - a_0 t - k. \tag{2.55}$$

If the initial wavefront Ω_0^k is given by the straight line $u_0 := \alpha x_1 + \beta x_2 - k = 0$, the wavefront at time t is Ω_t^k given by $u := \alpha x_1 + \beta x_2 - a_0 t - k = 0$. The straight line wavefront moves parallel to itself with velocity a_0.

Case 2 Next we take $\varphi_0 = \alpha x_1 + \beta x_2 + (\alpha x_1 + \beta x_2)^2$, i.e. we choose

$$u_0 = \alpha x_1 + \beta x_2 + (\alpha x_1 + \beta x_2)^2 - k, \quad \alpha, \beta = constants, \quad \alpha^2 + \beta^2 = 1. \tag{2.56}$$

A parametric representation of the Cauchy data at $\sigma = 0$ is

$$x_{10} = \eta_1, \quad x_{20} = \eta_2, \quad t_0 = 0 \quad u_0 = \alpha\eta_1 + \beta\eta_2 + (\alpha\eta_1 + \beta\eta_2)^2 - k. \tag{2.57}$$

Initial values of p_{10}, p_{20} and q_0 turn out to be

$$p_{10} = \alpha + 2\alpha(\alpha\eta_1 + \beta\eta_2), \quad p_{20} = \beta + 2\beta(\alpha\eta_1 + \beta\eta_2), \quad q_0 = -a_0\{1 + 2(\alpha\eta_1 + \beta\eta_2)\}. \tag{2.58}$$

Replacing σ by t in this case also, we write the characteristic equations and compatibility conditions as

$$\frac{dx_1}{dt} = a_0\frac{p_1}{|\mathbf{p}|}, \ \frac{dx_2}{dt} = a_0\frac{p_2}{|\mathbf{p}|}, \ \frac{dp_1}{dt} = 0, \ \frac{dp_2}{dt} = 0, \ \frac{dq}{dt} = 0, \ \frac{du}{dt} = a_0\sqrt{p_1^2 + p_2^2} + q. \tag{2.59}$$

Since the solution is generated by the Monge strips, we can use the PDE (2.49) to simplify the expressions on the right-hand side of the last equation of (2.59) so that $\frac{du}{dt} = 0$. Solving these equations with initial conditions (2.57) and (2.58), we get

$$x_1 = \eta_1 + a_0 \frac{p_{10}}{|\boldsymbol{p}_0|} t, \quad x_2 = \eta_2 + a_0 \frac{p_{20}}{|\boldsymbol{p}_0|} t, \quad p_1 = p_{10}, \quad p_2 = p_{20}, \quad q = q_0, \quad u = u_0, \tag{2.60}$$

where the terms on the right-hand side are functions of t, η_1 *and* η_2. Since $|\boldsymbol{p}_0| = 1+2(\alpha\eta_1+\beta\eta_2)$ and $\alpha p_{10}+\alpha p_{20} = 1+2(\alpha\eta_1+\beta\eta_2)$; $\alpha x_1+\beta x_2 = \alpha\eta_1+\beta\eta_2+a_0 t$.

Using the last result in $u = u_0 = \alpha\eta_1 + \beta\eta_2 + (\alpha\eta_1 + \beta\eta_2)^2 - k$, we get the solution as

$$u = (\alpha x_1+\beta x_2-a_0 t)+(\alpha x_1+\beta x_2-a_0 t)^2-k = (\alpha x_1+\beta x_2-a_0 t)(1+\alpha x_1+\beta x_2-a_0 t)-k. \tag{2.61}$$

In the **Case 1,** Ω_0^k consists a one-parameter family of parallel straight lines. The resulting wavefronts at time t are also straight lines displaced from the originals by distance $a_0 t$. In the **Case 2,** Ω_0^k consists of one-parameter family of curved lines (except for Ω_0^0) and Ω_t^k consists of similar displacement of each member of this family. It would be interesting to workout the wavefronts produced by the family $\Omega_0 : u_0 \equiv \alpha x_1 + \beta x_2 + k(\alpha x_1 + \beta x_2)^2 = 0$.

Remark 2.4.2 We have presented the theory of first-order PDEs briefly. It is based on the existence of characteristics curves in the $(\boldsymbol{x}, t)$-space. Along each of these characteristics, we derive a number of compatibility conditions, which are transport equations and which are sufficient to carry all necessary information from the datum curve in the Cauchy problem into a domain in which the solution is determined. Therefore, every first-order PDE is a hyperbolic equation.[4]

[4]A classification of equations into hyperbolic and other type of equations is done precisely on the same principle *'hyperbolicity of a higher order equation or a system of equations is due to the fact that it has sufficient number of families of characteristic curves (or bicharacteristics for equations in more than two independent variables), which carry all necessary information from the datum curve (or surface) to a point P not on the datum curve (or surface) to give the solution of a Cauchy problem at the point P'*.

Chapter 3
Scalar Conservation Law

Many important concepts of the theory of hyperbolic conservation laws can be explained with the help of a scalar conservation law[1] **in two independent variables**, which we take up in this chapter. Some of the results for a system of conservation laws and some others for conservation laws in multi-dimensions (i.e. in more than two independent variables) will be quoted very briefly whenever necessary in later chapters.

3.1 Conservation Law and Curves of Discontinuity

While considering the solution of the Eq. (2.32), namely $u_t + uu_x = 0$ in example (2.2.3), we saw that even for a smooth initial data the solution develops a singularity after a critical time t_c and the smooth solution of this problem cannot be continued. However, Eq. (2.32) models quite a few physical phenomena, where states represented by the function u becomes discontinuous after the time t_c and the discontinuous states of the phenomena persist for all time $t > t_c$. Hence, we need to generalize the notion of a solution to permit solutions, which are not necessarily C^1.

This generalization is simple, when we note that laws of physics are primarily stated not point wise but in integral forms, which remain valid also for discontinuous processes. A law for a change in time interval (t_1, t_2), $t_2 > t_1$ of a physical quantity $H(u(\xi, t))$ (which we call density) contained in a fixed space interval (x_1, x_2) is due to the net flux $F(u(x, t))$ of the quantity through the end points x_1 and x_2, assuming that there is no production of the quantity in the given interval. This is expressed in the form

$$\int_{x_1}^{x_2} H(u(\xi, t_2))d\xi - \int_{x_1}^{x_2} H(u(\xi, t_1))d\xi = \int_{t_1}^{t_2} \{F(u(x_1, t)) - F(u(x_2, t))\}dt, \quad (3.1)$$

[1] Also referred in the literature as single conservation law.

P. Prasad, *Propagation of Multidimensional Nonlinear Waves and Kinematical Conservation Laws*, Infosys Science Foundation Series,
https://doi.org/10.1007/978-981-10-7581-0_3

which holds for every fixed space interval (x_1, x_2) and for every time interval (t_1, t_2). This equation is meaningful even for a discontinuous function $u(x, t)$. A more restricted form of (3.1) is

$$\frac{d}{dt}\int_{x_1}^{x_2} H(u(\xi, t))d\xi = F(u(x_1, t)) - F(u(x_2, t)), \quad x_1, x_2 \text{ fixed.} \tag{3.2}$$

We symbolically denote the laws (3.1) or (3.2) by an equation in the form

$$\frac{\partial H(u)}{\partial t} + \frac{\partial F(u)}{\partial x} = 0 \tag{3.3}$$

and call it a **conservation law**. We assume that the two functions H and F are smooth functions of u. We can show that *for a C^1 function u, the above conservation laws are equivalent to the PDE*

$$H'(u)u_t + F'(u)u_x = 0. \tag{3.4}$$

Definition 3.1.1 A conservation law is an equation in a divergence form (3.3) and by this, we mean (3.1) or (3.2).

Definition 3.1.2 We define a **weak solution** of the conservation law (3.3) to be a bounded measurable function $u(x, t)$ which satisfies the integral form (3.1) or in particular (3.2).

This definition is the original one with which theory of shock wave started [14]. In all books in PDE today, a weak solution is defined as a distribution with the help of test functions, differently from that in the Definition 3.1.2, see [17, 47, 51], where one takes $u \in L^\infty_{loc}(D),\ D \in \mathbb{R}^2$. References [17, 51] contain a rigorous treatment of a system of conservation laws in one space variable. As mentioned earlier, the Ref. [42] explains all basic concepts in a very simple way for an equation when the flux function is convex. This book also has many examples to highlight basic properties.

A conservation law (one of infinity of conservation laws), from which the Burgers equation (2.32) i.e., $u_t + uu_x = 0$ can be derived for $u \in C^1$, is

$$\frac{\partial}{\partial t}(u) + \frac{\partial}{\partial x}\left(\frac{1}{2}u^2\right) = 0. \tag{3.5}$$

Jump relation across a curve of discontinuity in (x, t)-plane: Consider now a **weak** solution $u(x, t)$ of the conservation law (3.3) in (x, t)-plane such that $u(x, t)$ and its partial derivatives suffer discontinuities across a smooth isolated curve $\Omega : x = X(t)$ and u is continuously differentiable elsewhere. It is further assumed that the limiting values of u and its derivatives as we approach Ω from either side exist. The function $u(x, t)$ is a genuine solution of (3.4) in the left and right subdomains of the curve of discontinuity Ω. Let the fixed points x_1 and x_2 be so chosen that

$x_1 < X(t) < x_2$ for $t \in$ an open interval. Writing $\int_{x_1}^{x_2} H(u(\xi,t))d\xi = \int_{x_1}^{X(t)} H(u(\xi,t))d\xi + \int_{X(t)}^{x_2} H(u(\xi,t))d\xi$ in (3.2) we get

$$\int_{x_1}^{X(t)} H' u_t(\xi,t)d\xi + \int_{X(t)}^{x_2} H' u_t(\xi,t)d\xi + \dot{X}(t)\{H(u(X(t)-,t)) - H(u(X(t)+,t))\}$$

$$= \{F(u(x_1,t)) - F(u(x_2,t))\}.$$

The first two terms tend to zero as $x_1 \to X(t)-$ and $x_2 \to X(t)+$. Hence, taking the point x_1 on the left of $X(t)$ very close to it and the point x_2 on the right of $X(t)$ also very close to it, we get in the limit

$$\dot{X}(t)(H(u_\ell(t)) - H(u_r(t))) = F(u_\ell) - F(u_r) \tag{3.6}$$

where

$$u_\ell(t) = \lim_{x \to X(t)-} u(x,t) \quad and \quad u_r(t) = \lim_{x \to X(t)+} u(x,t). \tag{3.7}$$

Equation (3.6) gives the following expression for the velocity of propagation $\dot{X}$ of the discontinuity

$$\dot{X}(t) = [F]/[H] \tag{3.8}$$

where the symbol [ ] is defined by

$$[f] = f(u_\ell) - f(u_r). \tag{3.9}$$

The **jump relation** (3.6) or (3.8), connecting the speed of propagation $\dot{X}(t)$ of a discontinuity and the limiting values u_ℓ and u_r on the two sides of the discontinuity, is called **Rankine–Hugoniot condition** or simply **RH condition**. Such jump relations were first derived for a discontinuity in gas dynamics.

For a discontinuity with a non-zero jump, $u_\ell \neq u_r$. Hence, the jump relation (3.8) for the conservation laws (3.5) becomes

$$\dot{X}(t) = \frac{1}{2}(u_r + u_\ell). \tag{3.10}$$

Remark 3.1.3 If we take a conservation law $\frac{\partial}{\partial t}(u^n) + \frac{\partial}{\partial x}(\frac{n}{n+1}u^{n+1}) = 0,\ n \in \mathbb{N}$, the jump relation gives the jump velocity as $\dot{X}(t) = \frac{n}{n+1}(\sum_{i=0}^{n} u_r^{n-i} u_\ell^i)/(\sum_{i=0}^{n-1} u_r^{n-1-i} u_\ell^i)$. This, is different from (3.10) for $n > 1$, showing that different conservation laws leading to the same PDE (2.32) for a smooth solution are not equivalent for a weak solution.

3.2 Stability Consideration, Entropy Condition and Shocks

We have seen in Chap. 2 that a smooth solution of an initial value problem for the PDE (3.4) is unique. We shall show below that this is not so for a weak solution of a conservation law, in particular (3.5). Hence, in order to to get a unique solution we need to impose an additional **stability condition** *'an acceptable solution of an initial value problem for a conservation law is the one which is stable with respect to small changes in the initial data'*. Let us formulate an easily verifiable mathematical criteria, called **entropy condition**, for the uniqueness of the solution from some simple examples and show its relation to the stability condition.

Note: In the class of discontinuous solutions, we may also consider a discontinuous initial data.

Example 3.2.1 Let us consider an initial value problem for the Eq. (3.5) with a discontinuous initial data

$$u(x,0) = u_{01}(x) \equiv \begin{cases} 0\,, & x \le 0 \\ 1\,, & 0 < x. \end{cases} \tag{3.11}$$

This problem has an infinity of weak solutions depending on a parameter α satisfying $0 \le \alpha \le 1$:

$$u(x,t) = \begin{cases} 0 & , \quad x \le 0 \\ x/t & , \quad 0 < x \le \alpha t \\ \alpha & , \quad \alpha t < x \le \frac{1}{2}(1+\alpha)t \\ 1 & , \quad \frac{1}{2}(1+\alpha)t < x. \end{cases} \tag{3.12}$$

For $0 \le \alpha < 1$, all solutions are discontinuous on the line $x = \frac{1}{2}(1+\alpha)t$ and the RH condition is satisfied on the curves of discontinuity. For $\alpha = 1$, the above expression gives a continuous solution but not a genuine solution since it has discontinuous derivatives on the lines $x = 0$ and $x = 1$. This solution has been shown in Fig. 3.1(ii). The characteristic curves have been shown by broken lines. We note that for $\alpha \ne 0, 1$ there are four domains with different expressions of the solution: (i) $u = 0,\ x \le 0$; (ii) $u = \frac{x}{t},\ 0 < x \le \alpha t$; (iii) $u = \alpha,\ \alpha t < x \le \frac{1}{2}(1+\alpha)t$; and (iv) $u = 1,\ \frac{1}{2}(1+\alpha)t < x$.

The non-uniqueness in the solution arises because of the possibility of fitting a line of discontinuity of an arbitrary slope $\frac{1}{2}(1+\alpha)$ joining a constant state continuation ($u = \alpha$) of the **centred fan** $u = x/t$ on the left and the value $u = 1$ on the right. Another simple solution with initial condition (3.11) consists of three constant states $u = 0, 1/2$ and 1 separated by lines of discontinuity $x = (1/4)t$ and $x = (3/4)t$, across which the jump relation (3.10) is satisfied. **We notice that the value of u on the left of each of the discontinuities is smaller than the value of u on the right.**

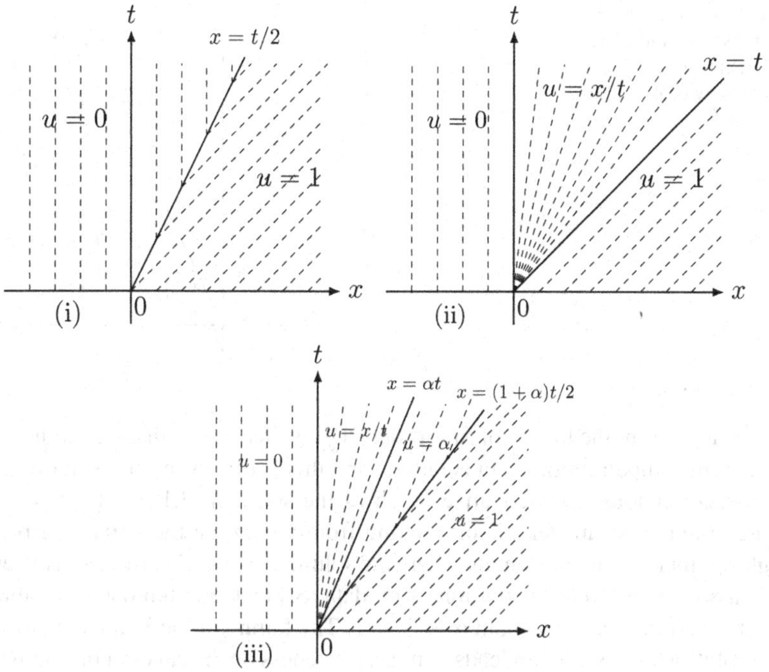

Fig. 3.1 Three figures show solution (3.12) for $\alpha = 0$, $\alpha = 1$ and $0 < \alpha < 1$. The characteristic curves have been shown by broken lines

Example 3.2.2 Instead of the initial data $u_{01}(x)$, if we take

$$u(x, 0) = u_{02}(x) \equiv \begin{cases} 1 & , \quad x \le 0 \\ 0 & , \quad x > 0 \end{cases} \tag{3.13}$$

we get a discontinuous solution of (3.5) shown in the Fig. 3.2:

$$u(x, t) = \begin{cases} 1 & , \quad x - \frac{1}{2}t \le 0 \\ 0 & , \quad x - \frac{1}{2}t > 0. \end{cases} \tag{3.14}$$

In this case, since the value of u in the state on the left is greater than that on the right, it is not possible to have a centred fan with centre at the origin and hence (3.14) is the only weak solution which has a curve of discontinuity through (0, 0).

The above two examples show that, in general, a discontinuous solution (i.e. a weak solution) of an initial value problem for (3.5) and hence, for a general conservation law (3.3) is not unique. Now, we need a mathematical principle characterizing a class of permissible solutions in which every initial value problem for the conservation law has a unique solution. We can deduce such a principle from the following

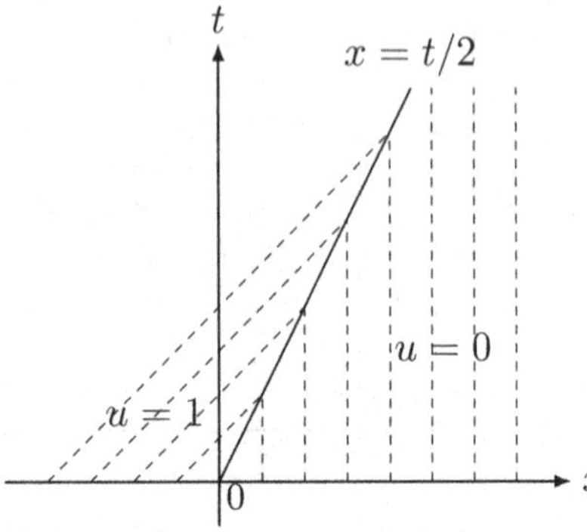

Fig. 3.2 There is a single curve of discontinuity. The characteristic curves have been shown by broken lines

consideration. From the theory of first-order PDE we have seen that a genuine solution satisfying smooth initial data is unique and this is true even for a solution with piecewise continuous first derivatives such as the solution (3.12) with $\alpha = 1$. In this case, a unique characteristic originating from a point on the initial line passes through a point (x, t) in the domain where the solution is to be determined. However, in the class of discontinuous solutions with data (3.11), the initial data is unable to control the solution in the domain $\alpha < \frac{x}{t} < 1$. The failure of the initial data to control the solution in this domain leads to non-uniqueness on the curve of discontinuity of the solution. When a discontinuity appears in the solution (3.12) for $0 \le \alpha < 1$, the characteristics starting from a point $(\frac{1}{2}(1+\alpha)t, t)$ diverge (into the domains on the two sides of it) as t increases, so that discontinuity could have been replaced by a continuous centred wave from this point onward. The situation is different when the initial data is (3.13), which gives rise to a situation in which characteristics starting from the points on the two sides of the point of discontinuity converge and start intersecting as t increases. The intersecting characteristics carry different values of of u. In this and all other such situations, a discontinuity must necessarily appear to prevent multivaluedness in the solution. Therefore, a discontinuity is permissible only if it prevents the intersection of the characteristics coming from the points of the initial line on the two sides of it, i.e.

$$u_r(t) < \dot{X}(t) < u_\ell(t).$$

If we accept this as a principle, Fig. 3.1 shows that the only admissible solution for initial data u_{01} is the continuous solution for $\alpha = 1$. A part of this solution between the lines $x = u_\ell t$ and $x = u_r t$ is called **centred expansion wave**, also known simply as **centred wave**. The solution (3.14) is also the only admissible solution for the initial data u_{02}.

We note that u is the **characteristic velocity** of the Burgers' equation (2.32), which is a differential form of (3.5). If we denote this characteristic velocity by $c(u)$, we can write the above inequality in the form

$$c(u_r)(t) < \dot{X}(t) < c(u_\ell)(t). \tag{3.15}$$

Now the principle for an admissible or acceptable discontinuity for (2.32) becomes (3.15).

Remark 3.2.3 In the discussion above, we note that in conservation law (3.5), the flux function $\frac{1}{2}u^2$ is convex. By replacing $H(U)$ by u, we write the general form of a scalar conservation law (3.3) as

$$\frac{\partial}{\partial t}(u) + \frac{\partial}{\partial x}(F(u)) = 0, \tag{3.16}$$

and we call it **scalar convex conservation law** if $F(u)$ is a convex function, i.e. $F'' > 0$. The characteristic velocity of the differential form of (3.16) is $c(u) = F'(u)$. Hence, the condition (3.15) again prevents multivaluedness in the solution. In fact, the condition (3.15), satisfied on every curve of discontinuity, is necessary and sufficient for a unique solution of the scalar convex conservation law.

A discontinuity in u which satisfies (3.15) is an **admissible discontinuity**. A weak solution, in which discontinuities in u are admissible, is **physically relevant**. Indeed, we have the following important theorem:

Theorem 3.2.4 *Let $u^\varepsilon(x,t)$ be the solution of the* ***viscous equation***

$$u_t + F(u)_x = \varepsilon u_{xx}, \quad F'' > 0, \tag{3.17}$$

with an initial value (1.5) for scalar u. Let $u(x,t) = \lim_{\varepsilon\to 0} u^\varepsilon(x,t)$ in L^1 norm. Then $u(x,t)$ is a weak solution of (3.16) and the condition (3.15) is satisfied on every curve of discontinuity.

Definition 3.2.5 An admissible discontinuity satisfying the stability criterion (3.15) is called a **shock**.

The term **shock** was first used for a compression discontinuity in gas dynamics, where an expansion discontinuity is ruled out by the second law of thermodynamics, which is consistent with the fact that the specific entropy of the fluid particles must increase after crossing the discontinuity.

Definition 3.2.6 The stability condition (3.15) is also called **entropy condition**. This is a particular case of **Lax entropy condition** for a system of conservation laws.

Theory of shocks for a system of conservation law is well developed [16, 17, 51] with various extensions of entropy conditions for a flux function F which is not necessarily convex. We do not go into these details but simply mention that one these conditions is **Liu entropy condition** and they are available in many books today [16].

3.3 Riemann Problem

One of the most important problems in the theory of conservation laws is the **Riemann problem**. In the Chap. 8, we shall use its solutions for KCL-based weakly nonlinear ray theory in gas dynamics to describe **elementary waves** on a wavefront and results of their interactions. Here, we solve this problem for the convex conservation law (3.16).

Definition 3.3.1 Let u_ℓ *and* u_r be constants, $u_\ell \neq u_r$. The Riemann problem for (3.16) is to find a weak solution satisfying the initial data

$$u(x,0) = \begin{cases} u_\ell \ , & x \leq 0 \\ u_r \ , & x > 0 \end{cases}, \tag{3.18}$$

Remark 3.3.2 Since we are considering a weak solution, we could have taken $u = u_\ell$ for $x < 1$ and $u = u_r$ for $x \geq 1$. Changing the initial value at a set of measure zero does not effect the solution, see solution with initial data (1.9.12) in [42].

Solution of the Riemann problem (3.16) **and** (3.18). Here F is a convex function. There are two cases giving two types of solutions, satisfying the entropy condition (3.15).
Case 1: When $u_\ell > u_r$

$$u(x,t) = \begin{cases} u_\ell \ , & x \leq St \\ u_r \ , & x > St \end{cases}, \tag{3.19}$$

where the shock velocity denoted by S is given by (3.8), i.e.

$$S := \dot{X} = \frac{F(u_\ell) - F(u_r)}{u_\ell - u_r}. \tag{3.20}$$

The two states u_ℓ and u_r are separated by a shock moving with the velocity S.
Case 2: When $u_\ell < u_r$, we define G to be inverse of the function F', i.e. $F' G(u) = u$, then

$$u(x,t) = \begin{cases} u_\ell \ , & x \leq tF'(u_\ell), \\ G(\frac{x}{t}) \ , & tF'(u_\ell) < x \leq tF'(u_r), \\ u_r \ , & x > tF'(u_r). \end{cases} \tag{3.21}$$

In this case, a centred wave appears which separates a constant state solution u_ℓ on the left from a constant sate solution u_r on the right. The centred wave has straight characteristics $x = tF'(u)$, $u_\ell \leq u \leq u_r$, all originating from the origin in (x,t)-plane.

3.4 Examples

We present now two examples briefly from [42], which highlight most important properties of solutions of a scalar convex conservation law in two independent variables, see Theorems 16.14 and 16.15 in [51].

Example 3.4.1 Let B be a constant > 0. Take an initial data

$$u_0(x) = \begin{cases} B \ , & -1 < x \le 1, \\ 0 \ , & x \le -1 \text{ and } x > 1. \end{cases} \tag{3.22}$$

The solution of (3.5) with this initial value has two distinct representations in two different time intervals:
(i) $0 < t \le \frac{4}{B}$.

$$u(x,t) = \begin{cases} 0 \ , & x \le -1, \\ \frac{x+1}{t} \ , & -1 < x \le -1 + Bt, \\ \frac{1}{2}A \ , & -1 + Bt < x \le 1 + \frac{B}{2}t, \\ 0 \ , & 1 < x, \end{cases} \tag{3.23}$$

which has been shown in the Fig. 3.3 in (x, u)-plane. We find that in the (x, t)-plane there is a centred wave in the wedged shape region $-1 < x \le -1 + Bt$ and a shock along the curve $x = 1 + \frac{1}{2}Bt$. At the time $t = \frac{4}{B}$ the leading front of the centred wave overtakes the shock at $x = 3$. After this time, the shock interacts with the centred wave.
(ii) $t \ge \frac{4}{B}$.

We get the shock path $x = X(t)$ by solving

$$\frac{dX}{dt} = \frac{1}{2}(u_\ell(X(t)) + u_r(X(t)) = \frac{X+1}{2t}, \quad X\left(t = \frac{4}{B}\right) = 3$$

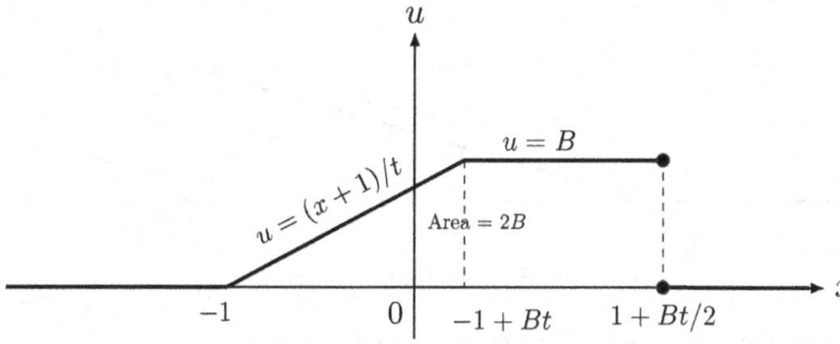

Fig. 3.3 Graph of the solution with initial value (3.22) valid in time interval $0 < t \le \frac{4}{B}$

and we get

$$X(t) = -1 + \sqrt{4Bt}. \tag{3.24}$$

At $x = X(t)$ the shock strength $u_\ell - u_r$ is given by

$$u = \frac{x+1}{t}|_{x=X(t)} = \sqrt{\frac{4B}{t}}. \tag{3.25}$$

We find that the pulse in (x, u)-plane has a triangular shape whose base is spread over a distance $\sqrt{4Bt}$ and whose height is $\sqrt{\frac{4B}{t}}$ (Fig. 3.4). The total area of the pulse $2B$ remains constant with time t and it equals to the area of the initial pulse given by (3.22).

The total area of the pulse in the above example remains constant as the pulse evolves agrees with the property of the conservation law (2.32) '*when the initial data of u vanishes outside a closed bounded interval of x-axis,* $\int_{-\infty}^{\infty} u(\xi, t)d\xi$ *is independent of t*'. Figure 3.4 gives the limiting form of the shape of the graph of any solution for which the initial data $\varphi(x)$ is positive everywhere and is of compact support.

Problem Work out the Example 3.4.1 with $B < 0$ and draw the figures corresponding to the Figs. 3.3 and 3.4.

Note that the Burgers' equation (2.32) remains invariant under a simultaneous change the sign of u and x. Hence, if we change the sign of B in (3.22), all results in the Example 3.4.1 can be easily worked out. There will be a centred rarefaction wave with centre at $(1, 0)$ with a negative value u in it and a shock with negative strength will start moving in negative x-direction from the position $x = -1$. The figures similar to Figs. 3.3 and 3.4 can be easily drawn.

Example 3.4.2 Consider an initial data

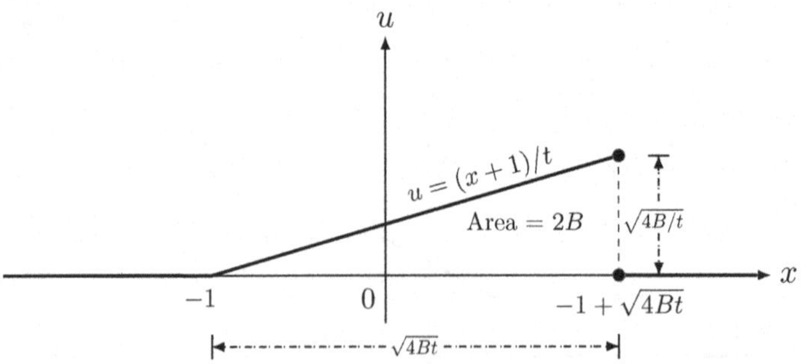

Fig. 3.4 Graph of the solution with initial condition (3.22) valid from $t > \frac{4}{B}$

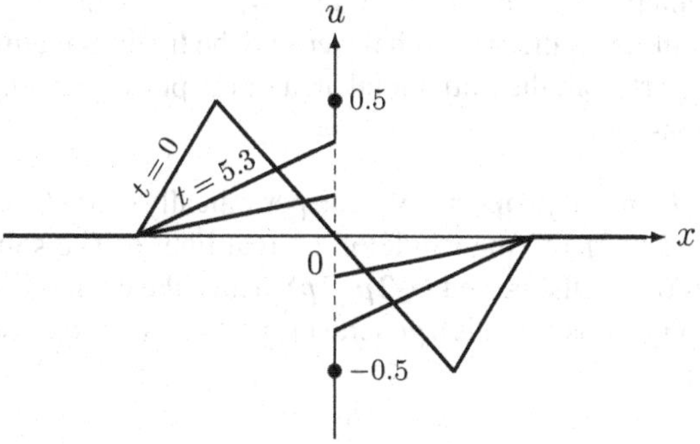

Fig. 3.5 Initial pulse, given by (3.26), is drawn for $p = \frac{1}{2}$ and $a = \frac{1}{2}$. The solution develops a shock at time $t = \frac{p}{a}$ at the origin, which decays to zero with time as $O(\frac{1}{t})$

$$\varphi(x) = \begin{cases} 0 \ , & -\infty < x \leq -2p, \\ a(x+2p)/p \ , & -2p < x \leq -p, \\ -ax/p \ , & -p < x \leq p, \\ a(x-2p)/p \ , & p < x \leq 2p, \\ 0 \ , & 2p < x < \infty. \end{cases} \tag{3.26}$$

where $p > 0$ and $a > 0$. This initial data has been shown graphically in Fig. 3.5.

The solution of the conservation law (3.5) with initial data (3.26) remains continuous for $0 < t < \frac{p}{a}$. At $t = \frac{p}{a}$, a shock appears at the origin with $u_\ell = a$ and $u_r = -a$. According to the jump relation, this shock does not move away from the origin but its amplitude decays. The solution in the interval $-2p < x \leq -p + at$ for $t \leq \frac{p}{a}$ and in $-2p < x < 0$ for $t > \frac{p}{a}$ are given by

$$u(x,t) = \frac{x+2p}{t+\frac{p}{a}}. \tag{3.27}$$

Since $u_r(t) = -u_\ell(t)$, from (3.27) it follows that the shock strength at the origin is

$$u_\ell - u_r = 4p/(t + \frac{p}{a}) \tag{3.28}$$

showing that, unlike all previous example where the shock strength decays as $0\left(\frac{1}{\sqrt{t}}\right)$, in this case, it decays as $0\left(\frac{1}{t}\right)$. The asymptotic solution as $t \to \infty$, retaining only the first term is

$$u(x,t) = \begin{cases} 0 \ , & -\infty < x \leq -2p, \\ (x+2p)/t \ , & -2p < x \leq 0, \\ (x-2p)/t \ , & 0 < x \leq 2p, \\ 0 \ , & 2p < x < \infty, \end{cases} \tag{3.29}$$

which has a shock of strength $4p/t$. It is interesting to note that the asymptotic solution is the same whatever may be the initial amplitude except that it remembers (apart from the initial total area of the pulse) the length $4p$ of the interval where it is non-zero.

A General property: We can periodically extend the initial data (3.26) on the interval $(-2p, 2p)$ to the whole of the real line $\mathbb{R}$. The solution in each period is the same as that in the period $(-2p, 2p)$. Thus, the example 3.4.2 shows that the decay of the solution as $O(\frac{1}{t})$ is a general property of a periodic solution of a convex conservation law.

3.5 Nonlinear Wavefront and Shock Front

The distinction between a nonlinear wavefront and shock front has been discussed in detail in [42]. We quote some sentences with a little modification from this reference.

'Solution $u = u_0(\xi)$, $\xi = x - ct$ of the PDE $u_t + cu_x = 0$, $c =$ constant, satisfying the initial condition $u(x, 0) = u_0(x)$ represents a pulse which is generated by one parameter family of wavefronts represented by $x = ct + \xi$, where $\xi \in \mathbb{R}$ is the parameter. u_0 need not be $C^1(\mathbb{R})$ in which case, u is a suitably defined generalized or weak solution. If u_0 is continuous, we get a continuous pulse in which the amplitude varies continuously across all wavefronts, each one of which moves with the same velocity c. If u_0 has a discontinuity at a point x_0, we get a wavefront $x = ct + x_0$ across which the amplitude is discontinuous but this wavefront also moves with the same velocity.

For the nonlinear conservation law (3.5), a distinction has to be made between a one parameter family of wavefronts $x = tu_0(\xi) + \xi$ forming a continuous pulse given by $u = u_0(x - ut)$ and a discontinuous front, i.e. a shock front. A shock is also a wavefront, if we give a general interpretation of the word 'wavefront' but a wavefront (across which the amplitude varies continuously) and a shock front have different **kinematical and dynamical** properties. By kinematical property, we mean the eikonal equation governing the evolution of the front and by dynamical property we mean the transport equations of the amplitude and its derivatives as we move along the front. Kinematics and dynamics get coupled for a nonlinear wavefront and a shock front (see Sects. 7.2–7.4).

In this monograph, we shall make a clear distinction between the use of the words *nonlinear wavefront* and *shock front*.

Definition 3.5.1 Consider an entropy satisfying solution of a conservation law representing a pulse in a genuinely nonlinear characteristic field. When the amplitude of the pulse varies continuously across a front, it will be called a **nonlinear wavefront** and when the amplitude is discontinuous across it, it will be called a **shock front**.

We note in Examples 3.4.1 and 3.4.2 that for the conservation law (3.5) the shock interacts with the nonlinear waves ahead of it and behind it and *swallows* them gradually so that information on an interval of increasing length in initial data gets lost.

Distinction between a linear wavefront, a nonlinear wavefront and a shock front is very easily understood in the case of 1-D or plane fronts. This becomes complex for waves in multi-dimensions due to the convergence or divergence of rays and due to nonlinear waves which propagate on the fronts themselves.'

Chapter 4
Hyperbolic System of PDEs and Conservation Laws

Before we take up a discussion of a theory of a hyperbolic system of first order of equations, we present kinematics of a moving surface Ω_t (in particular a wavefront) in d-dimensional space. In this, we shall discuss the evolution of Ω_t with the help of rays. We shall see that not every vector χ qualifies as a **ray velocity** [43].

4.1 Ray Equations in a General System in *d*-D Space

Waves involve the transfer of energy from one part of a medium to another part, usually without transfer of material particles [10]. When we take such a general definition of waves, we may not be in a position to identify some propagating surfaces which we shall like to call **wavefronts** Ω_t. Identification of a wavefront[1] Ω_t requires an approximation: there is a more rapid change in the state of the medium as we cross the wavefront transversely compared to more gradual changes in the state, which is already present prior to the onset of the wave, or when we move along the wavefront. Thus, when we encounter a wave we can see a **short wavelength** variation in the state of the system at a given time or a **high-frequency** variation with respect to time at a fixed point. Identification of a wavefront Ω_t requires finite speed C of propagation of Ω_t. Let Ω_t be represented by

$$\Omega_t : \varphi(\mathbf{x}, t) = 0, \quad \mathbf{x} \in I\!R^d, t \in I\!R \tag{4.1}$$

then

$$C = -\varphi_t / \mid \nabla\varphi \mid . \tag{4.2}$$

Evolution of Ω_t takes place with the help of a ray velocity χ which can be obtained only when the dynamics of the curve Ω_t is known, i.e. the nature of the medium in which Ω_t propagates. For example, when Ω_t is a crest line of a curved solitary

[1] A mathematical definition of a wavefront is available in [42], Sect. 3.2.1, originally given in [36].

P. Prasad, *Propagation of Multidimensional Nonlinear Waves and Kinematical Conservation Laws*, Infosys Science Foundation Series,
https://doi.org/10.1007/978-981-10-7581-0_4

wave on shallow water, χ is given in an approximation to governing elliptic system of equations of water and boundary conditions on the surface of the water and the bottom surface [6].

The ray velocity χ at any point $\boldsymbol{x}$ of Ω_t depends also on the unit normal $\boldsymbol{n}$ of Ω_t at $\boldsymbol{x}$. Thus, $\chi = \chi(\boldsymbol{x}, t, \mathbf{n})$. The velocity C of Ω_t is the normal component of χ, i.e.

$$C = \langle \boldsymbol{n}, \chi \rangle, \qquad \boldsymbol{n} = \nabla\varphi / \mid \nabla\varphi \mid. \tag{4.3}$$

Using (4.2) and (4.3) we get the eikonal equation[2]

$$\varphi_t + \langle \chi, \nabla\varphi \rangle = 0 \tag{4.4}$$

which is a first-order nonlinear partial differential equation for φ describing the evolution of Ω_t.

Theorem 4.1.1 From [43]. *In order that the vector* $\chi(\mathbf{x}, t, \mathbf{n}) = (\chi_1, \chi_2, \dots, \chi_d)$ *qualifies to be a ray velocity, it must satisfy a consistency condition*

$$n_\beta n_\gamma \left(n_\beta \frac{\partial}{\partial n_\alpha} - n_\alpha \frac{\partial}{\partial n_\beta} \right) \chi_\gamma = 0, \quad \text{for each } \alpha = 1, 2, \cdots, d, \tag{4.5}$$

where summation convention has been used on repeated subscripts.

Proof Derivation of this condition is simple. Equation (4.4) is a nonlinear first-order equation written explicitly in the form $\varphi_t + \langle \chi(\boldsymbol{x}, t, \nabla\varphi / \mid \nabla\varphi \mid), \nabla\varphi \rangle = 0$. The Charpit equations of this first-order equation reduce to Hamilton's canonical equations for x_α and φ_{x_α}. In the equation for x_α we convert the derivatives with respect to φ_{x_α} into derivatives with respect to n_α using the second relation in (4.3). The equation for x_α turns out to be

$$\frac{dx_\alpha}{dt} = \chi_\alpha + n_\beta n_\gamma \left(n_\beta \frac{\partial}{\partial n_\alpha} - n_\alpha \frac{\partial}{\partial n_\beta} \right) \chi_\gamma.$$

Due to the presence of the second term on the right-hand side, this equation is inconsistent with the statement that χ_α is a ray velocity i.e., $\frac{dx_\alpha}{dt} = \chi_\alpha$ and hence the second term must vanish. The theorem is proved. The equation for x_α now takes the form (4.6) written below. ■

Now we take up the second part of the Hamilton's canonical equations, namely the equation for φ_{x_α}. We again use the second relation in (4.3) and convert the equation for φ_{x_α} into an equation for n_α and get (4.7) written below.

Thus, when condition (4.5) is satisfied, we have derived the ray equations from the eikonal (4.4) in the form

[2] Equation (4.2) gives another form of the eikonal equation, which is $\varphi_t + C \mid \nabla\varphi \mid = 0$. We can get (4.4) from it using $C = \langle \boldsymbol{n}, \chi \rangle$.

$$\frac{dx_\alpha}{dt} = \chi_\alpha, \tag{4.6}$$

$$\frac{dn_\alpha}{dt} = -n_\beta n_\gamma \left(\frac{\partial}{\partial \eta_\beta^\alpha}\right) \chi_\gamma \equiv \psi_\alpha, \text{ say} \tag{4.7}$$

where

$$\frac{\partial}{\partial \eta_\beta^\alpha} = n_\beta \frac{\partial}{\partial x_\alpha} - n_\alpha \frac{\partial}{\partial x_\beta}. \tag{4.8}$$

The derivatives $\frac{\partial}{\partial \eta_\beta^\alpha}$ represent tangential derivatives on the surface Ω_t. Since $|\boldsymbol{n}| = 1$, only $d-1$ equations in (4.7) are independent.

We may be tempted to ask a question: does a ray defined by (4.6) and (4.7) starting from one point $P_0 \in I\!R^d$ to another point $P_1 \in I\!R^d$ chooses a path such that the time of transit is stationary with respect to small variations in the path? Our question is about equivalence of rays given by (4.6) and (4.7) to the rays satisfying Fermat's Principle. It is easy to see this equivalence [42] for an isotropic propagation of a wavefront Ω_t, for which

$$\boldsymbol{\chi} = \boldsymbol{n}\, C \tag{4.9}$$

where C is independent of $\boldsymbol{n}$. In this case (4.5) is automatically satisfied. It will be interesting to prove this equivalence for a general ray velocity $\boldsymbol{\chi}$ satisfying (4.5). Some results on this equivalence are discussed in [44].

An important class of waves, called **hyperbolic waves**, appear in a medium governed by a hyperbolic system of partial differential equations. In this case, every signal in the medium propagates with a finite speed in a very strict sense: if the state of the system is perturbed at any time in a closed bounded domain, then the effect of the perturbation at any later time is not felt outside another closed bounded domain. The signal in such a domain may travel with a shock ray velocity $\boldsymbol{\chi}$ when the governing hyperbolic system is derived from a system of conservation laws. In the case of a curved solitary wave mentioned above, the **crest line** does represent a wavefront propagating with a **finite speed** but the solitary wave is not a hyperbolic wave [6]. It is governed by the KdV equation.

4.1.1 Wavefront Construction with the Help of Rays

Let Ω_{t_0} be the position of the wavefront at time t_0 and P_{t_0} (with coordinates $\boldsymbol{x}_{t_0}$) be a point on it. From the geometry of Ω_{t_0}, we calculate its normal direction $\boldsymbol{n}_{t_0}$ at P_{t_0}. Given the ray velocity $\boldsymbol{\chi}$, we solve the equations (4.6) and (4.7) and construct the ray starting from the point P_{t_0} and let this ray reach a point P_t ($\boldsymbol{x}_t$) at time t. The wavefront Ω_t is obtained as the locus of the point P_t ($\boldsymbol{x}_t$) as P_{t_0} takes different positions on Ω_{t_0}.

4.2 Hyperbolic System of First-Order PDE

In this section, we aim to deal mainly the theory of a hyperbolic system (1.4) in more than two independent variables. But we need to introduce first some concepts which we deal very briefly for a first-order system of n quasilinear PDE (1.3) in two independent variables x and t, i.e. a hyperbolic system in one space variable. As in the case of scalar first-order PDE, for a quasilinear system also we start with a known solution $\boldsymbol{u}(\boldsymbol{x}, t)$.

4.2.1 Hyperbolic System in One Space Variable

Consider the system of quasilinear equations (1.3).

Definition 4.2.1 The system is defined to be hyperbolic if **(i)** the matrix B has n real eigenvalues relative to the matrix A, i.e. the n roots of the equation in λ

$$\det(B - \lambda A) = 0 \tag{4.10}$$

are real and **(ii)** the dimension of the eigenspace of each eigenvalue is equal to its algebraic multiplicity.

We denote the eigenvalues by $c_1, c_2, \ldots, c_n$ and assume that

$$c_1 \leq c_2 \leq c_3 \leq \cdots \leq c_n. \tag{4.11}$$

Let $\boldsymbol{\ell}^{(i)}$ and $\boldsymbol{r}^{(i)}$ respectively be the left and right eigenvectors corresponding to c_i, i.e.

$$\boldsymbol{\ell}^{(i)}(B - c_i A) = 0 \text{ and } (B - c_i A)\boldsymbol{r}^{(i)} = 0, \; i = 1, 2, \ldots, n. \tag{4.12}$$

Characteristic curves corresponding to the eigenvalue c_i are curves in the (x, t)-plane given by the equation

$$\frac{dx}{dt} = c_i \tag{4.13}$$

Multiplying (1.3) by $\boldsymbol{\ell}^{(i)}$, and using (4.12) we get the following **compatibility condition** along a characteristic curve of the **ith family**

$$\boldsymbol{\ell}^{(i)} A(\partial_t + c_i \partial_x)\boldsymbol{u} + \boldsymbol{\ell}^{(i)} \boldsymbol{C} = 0, \tag{4.14}$$

which means

$$\boldsymbol{\ell}^{(i)} A \frac{d\boldsymbol{u}}{dt} + \boldsymbol{\ell}^{(i)} \boldsymbol{C} = 0 \text{ along } \frac{dx}{dt} = c_i. \tag{4.15}$$

Note that if an eigenvalue c_i appears a number of times (say p) in (4.11), i.e. c_i is a multiple eigenvalue of multiplicity p, there exist p linearly independent eigenvectors corresponding to this eigenvalue and we get p independent compatibility conditions corresponding to it.

We note that for a hyperbolic system it is possible to choose the left eigenvectors $\boldsymbol{\ell}^{(i)}$ and the right eigenvectors $\boldsymbol{r}^{(i)}$ in such a way that

$$\boldsymbol{\ell}^{(i)} A \boldsymbol{r}^{(j)} \begin{cases} = 0 \text{ for } i \neq j \\ = 1 \text{ for } i = j \end{cases} \tag{4.16}$$

For a quasilinear system, though the compatibility condition (4.15) is not linear in $\boldsymbol{u}$ since $\boldsymbol{\ell}$ depends on $\boldsymbol{u}$, it gives the ith compatibility condition in which a linear combination of only the rates of change of the n components of $\boldsymbol{u}$ along the ith characteristic appears. All these n transport equations together are equivalent to the original system. However, in the case of a linear system, we can get a canonical system of n equations for **characteristic variables** $\boldsymbol{w} = (w_1, w_2 \dots w_n)$. For this, we use the transformation

$$\boldsymbol{u} = \sum_{j=1}^{n} \boldsymbol{r}^{(j)} w_j \tag{4.17}$$

in (4.15) and use (4.16) (for a detailed discussion of its role, see [47]). The ith equation in this canonical system for $\boldsymbol{w}$ gives the time rate of change of the ith characteristic variable w_i along the characteristic curves of the ith **characteristic family** of (4.13). This implies that any solution of the linear hyperbolic system consists of n parts $w_1, w_2, \dots, w_n$. The part w_i propagates along the x-axis with a velocity equal to the ith eigenvalue c_i and the mutual interaction between these parts takes place only through the source terms in the canonical system.

The above beautiful result of a linear system does not hold for a quasiliear system but gets modified into another type of beautiful result the basis of which is the existence of simple waves, which was first studied by Riemann in his work on equations of a compressible fluid. Riemann's theory also leads to a new concept, **genuine nonlinearity**, first defined by Lax [28]. A very detailed discussion of simple waves is available in [42] not only in one space dimension but also in multi-dimensions and this reference also has examples.

4.2.2 *Simple Wave in One Space Dimension*

Consider a hyperbolic system of n equations

$$A(\boldsymbol{u})\frac{\partial \boldsymbol{u}}{\partial t} + B(\boldsymbol{u})\frac{\partial \boldsymbol{u}}{\partial x} = 0, \tag{4.18}$$

which is called a **reducible system** since A and B depend only on $\boldsymbol{u}$ and $\boldsymbol{C} = 0$ in (1.3).

Definition 4.2.2 A simple wave solution of this system is a genuine solution in a domain S such that all components u_i of $\boldsymbol{u}$ can be expressed in terms of a single function $w(x,t) \in C^1(S)$:

$$\boldsymbol{u}(x,t) = \boldsymbol{U}(w(x,t)). \tag{4.19}$$

Substituting (4.19) in (4.18) and assuming that $w_x \neq 0$, we get

$$[B(\boldsymbol{U}) - (w_t/w_x)A(\boldsymbol{U})]\frac{d\boldsymbol{U}}{dw} = 0. \tag{4.20}$$

For a nontrivial solution of $\frac{d\boldsymbol{u}}{dw}$ we require that $-w_t/w_x$ is equal to an eigenvalue c of (4.18). Therefore, w satisfies

$$w_t + c(\boldsymbol{U}(w))w_x = 0 \ \text{ in } S. \tag{4.21}$$

Assumption 4.2.3 We assume in this section that the real eigenvalue c is distinct.

The Eq. (4.21) implies that in S, w=constant i.e., $\boldsymbol{U}$ = constant along lines $x - c(\boldsymbol{U}(w))t$ = constant. The straight lines $x - c(\boldsymbol{U}(w))t$ = constant, are characteristic curves of the system (4.18) corresponding to the eigenvalue c. The derivative $\frac{d\boldsymbol{U}}{dw}$ is parallel to the corresponding right eigenvector, i.e.

$$\frac{d\boldsymbol{U}}{dw} = \boldsymbol{r}(\boldsymbol{U}). \tag{4.22}$$

Integrating this autonomous system of n ordinary differential equations we get

$$\boldsymbol{U} = \boldsymbol{U}(w, \boldsymbol{\pi}), \tag{4.23}$$

with $n-1$ arbitrary quantities $\boldsymbol{\pi} = (\pi_1, \pi_2, \dots, \pi_n - 1)$, which are independent of w. Note that $\boldsymbol{\pi}$ remains constant in S. Since w and $\pi_1, \dots, \pi_{n-1}$ are independent, the mapping between $(w, \pi_1, \dots, \pi_{n-1})$ and $(u_1, \dots, u_n)$ is one to one locally. We can express $w, \pi_1, \dots, \pi_{n-1}$ as functions of $u_1, \dots, u_n$:

$$w = w(\boldsymbol{u}), \quad \pi_\delta = \pi_\delta(\boldsymbol{u}), \quad \delta = 1, 2, \dots, n-1. \tag{4.24}$$

Since $\frac{\partial \pi_\delta}{\partial w} = 0$, using (4.22) in $\frac{\partial \pi_\delta}{\partial w} = \sum_{i=1}^{n} \frac{\partial \pi_\delta}{\partial u_i}\frac{\partial u_i}{\partial w}$ we get for each π_δ

$$\langle \boldsymbol{\nabla}_{\boldsymbol{u}} \pi_\delta, \boldsymbol{r} \rangle = 0, \quad \delta = 1, 2, \dots, n-1. \tag{4.25}$$

This is a first-order quasilinear partial differential equation for π

$$r_1(\boldsymbol{u})\pi_{u_1} + r_2(\boldsymbol{u})\pi_{u_2} + \dots + r_n(\boldsymbol{u})\pi_{u_n} = 0 \tag{4.26}$$

which has a set of $n-1$ independent solutions $\boldsymbol{\pi}(\boldsymbol{u}) = (\pi_1(\boldsymbol{u}), \ldots, \pi_{n-1}(\boldsymbol{u}))$. This leads to the definition of a new set of variables [28]:

Definition 4.2.4 A solution of (4.26) is called a **Riemann invariant** of the characteristic field associated with the eigenvalue $c(\boldsymbol{u})$.

Thus, there are $n-1$ independent Riemann invariants of the eigenvalue $c(\boldsymbol{u})$.

Definition 4.2.5 If $c(\boldsymbol{u})$ is the kth eigenvalue of the hyperbolic system (1.3), then the simple wave, in which the characteristics $\frac{dx}{dt} = c(\boldsymbol{u}(w))$ are straight lines, is said to be a **simple wave of kth characteristic field**.

Remark 4.2.6 In the simple wave of kth characteristic field, all $n-1$ Riemann invariants π_δ of kth characteristic field are constant.

We are now in a position to extend the notion of a **characteristic variable** appearing in (4.17) for linear hyperbolic system to that for a quasilinear system.

Definition 4.2.7 A function $w(\boldsymbol{u})$ such that

$$\frac{\partial(w, \pi_1, \pi_2, \ldots, \pi_{n-1})}{\partial(u_1, u_2, \ldots, u_n)} \neq 0 \tag{4.27}$$

in a domain of $\boldsymbol{u}$-space, is called **characteristic variable** of the eigenvalue $c(\boldsymbol{u})$.

In a simple wave associated with the eigenvalue c, all Riemann invariants $\pi_1, \pi_2, \ldots, \pi_{n-1}$ have constant values everywhere but the characteristic variable w has different constant values on different members of the characteristic curves $x - c(\boldsymbol{U}(w))t =$ constant, which are straight lines. In the simple wave, we write $c(\boldsymbol{U}(w))$ simply as $c(w)$. The equation governing the evolution of the characteristic variable w in the simple wave is a scalar first-order quasilinear partial differential equation (4.21), which we write as

$$w_t + c(w)w_x = 0. \tag{4.28}$$

Consider a constant value w_0 of w along a particular characteristic curve of in the simple wave. In the neighbourhood of this characteristic curve, we have

$$c(\boldsymbol{U}(w)) = c_0 + \left\{ \langle \nabla_{\boldsymbol{u}}\, c, \frac{d\boldsymbol{U}}{dw} \rangle \right\}_0 (w - w_0) + 0\{(w - w_0)^2\}.$$

Using the result that $\frac{d\boldsymbol{U}}{dw}$ is parallel to $\boldsymbol{r}$, we find that, to the first order in $(w - w_0)$, the characteristic velocity c in (4.28) varies linearly with w if $\langle \nabla_{\boldsymbol{u}} c, \boldsymbol{r} \rangle \neq 0$.

Definition 4.2.8 If

$$\langle \nabla_{\boldsymbol{u}} c, \boldsymbol{r} \rangle \neq 0, \quad \forall \boldsymbol{u} \in D_u \tag{4.29}$$

where D_u is a domain in $(u_1, u_2, \ldots, u_n)$-space, we say that the characteristic field under consideration is **genuinely nonlinear** in D_u.

When

$$\langle \nabla_u c, \boldsymbol{r} \rangle = 0, \quad \forall \boldsymbol{u} \in D_u$$

the characteristic field is called **linearly degenerate**.

The simplest example of an equation in which the only characteristic field is genuinely nonlinear is (2.32). An important example of a linearly degenerate characteristic field appears in the Euler equations (see (4.85)). Another example of a linearly degenerate characteristic field is in our main theme, i.e. theory of KCL (see Sects. 8.4, 8.8; and Theorems 10.1.3 and 10.2.1).

The following two results, which are simple to prove ([14] for a system of two equations; and [29, 47] for a general system), show that a simple wave occurs in a most natural way. We state these results without proof.

Theorem 4.2.9 *If a section of a kth characteristic carries a constant value of* $\boldsymbol{u}$*, then in a region adjacent to this section, the solution is either a constant state or a simple wave of kth family.*

Theorem 4.2.10 *The solution in a region adjacent to a region of the constant state is a simple wave solution.*

Now, we proceed to discuss a special class of simple wave solutions, namely the **centred simple wave**, more appropriately called a **centred rarefaction wave.**[3]

Definition 4.2.11 A centred simple wave is a genuine solution of (4.18) for $t > t_0$, in which $\boldsymbol{u}$ depends only on the ratio $(x - x_0)/(t - t_0)$; (x_0, t_0) being the centre of the wave.

Choosing the variable w in (4.18) to be $(x - x_0)/(t - t_0)$, we find that such a solution is a simple wave with one of the n families of characteristics, say kth, being straight lines represented by $(x - x_0)/(t - t_0) = \text{constant}$, all of which pass through the point (x_0, t_0). These characteristics diverge from one another as t increases. Now we state another theorem, which is very easy to prove.

Theorem 4.2.12 *Let a constant state* $\boldsymbol{u}_\ell$ *on left be connected to another constant state* $\boldsymbol{u}_r$ *on right by a centred simple wave of the kth family, then*

$$c_k(\boldsymbol{u}_\ell) < c_k(\boldsymbol{u}_r). \tag{4.30}$$

The inequality (4.30) is of great importance to us. Let us explain it with a particular choice of a parameter δ defined by:

$$\delta = c_k(\boldsymbol{u}_r) - c_k(\boldsymbol{u}_\ell). \tag{4.31}$$

In general, in a simple wave (not necessarily centred), δ can take both positive and negative values. However, in a centred wave the inequality (4.30) permits only positive values of δ since characteristics diverge as t increases.

[3]The name 'rarefaction wave' is borrowed from gas dynamics, where the density of fluid decreases as fluid elements cross a centred simple wave.

4.3 Hyperbolic System in Multi-dimensions

We quickly review some results on a hyperbolic system in multi-dimensions (by this we mean **multi-space-dimensions**). For many details see [13], [47] and [42]. We first take up an example: the wave equation and introduce the concepts of a characteristic surface, bicharacteristic equations and a ray.

4.3.1 The Wave Equation in d(>1) Space Dimensions

The wave equation in d-space dimension for the function $\boldsymbol{u}(\boldsymbol{x}, t)$ is

$$u_{tt} - a_0^2(u_{x_1x_1} + \cdots + u_{x_dx_d}) = 0, \quad a_0 = \text{ constant } > 0. \tag{4.32}$$

The generalization of a characteristic curve discussed in the last section is a **characteristic surface** Ω in $(\boldsymbol{x}, t)$-space. If we represent a characteristic surface Ω of (4.32) by $\varphi(\boldsymbol{x}, t) = \alpha$, α = constant, then $\varphi(\boldsymbol{x}, t)$ satisfies the characteristic partial differential equation

$$Q(\nabla\varphi, \varphi_t) \equiv \varphi_t^2 - a_0^2(\varphi_{x_1}^2 + \cdots + \varphi_{x_d}^2) = 0, \ \textit{where} \ \nabla\varphi = (\varphi_{x_1}, \ldots, \varphi_{x_d}). \tag{4.33}$$

The most important solution of (4.33) is

$$\varphi \equiv (t - t_0) \pm \frac{1}{a_0}|\boldsymbol{x} - \boldsymbol{x}_0| = 0. \tag{4.34}$$

It represents a **characteristic conoid** in $(\boldsymbol{x}, t)$-space with its vertex at the point $P_0(\boldsymbol{x}_0, t_0)$. In (4.34) $+$ and $-$ signs correspond respectively to the lower and upper branches of the conoid in space–time from P_0. For simplicity of discussion, we take $t_0 > 0$. Intersection of the conoid by the hyperplane $t = 0$ is a sphere

$$S_0 : |\boldsymbol{x} - \boldsymbol{x}_0|^2 = a_0^2(t_0)^2 \tag{4.35}$$

in $\boldsymbol{x}$-space.

Space-like surface: Consider a solution u of the wave equation (4.32). A d-dimensional surface R in space–time is said to be **space-like** if the value of u at any point P on R does not influence the solution u at other points of R.

An example of a space-like surface for the wave equation (4.32) is a hyperplane t = constant. Any other plane

$$v(t - t_0) - \langle \boldsymbol{n}, (\boldsymbol{x} - \boldsymbol{x}_0)\rangle = 0, \quad (v, \boldsymbol{n} \text{ are constant } \neq 0) \tag{4.36}$$

through $P_0(\boldsymbol{x}_0, t_0)$ such that it intersects the characteristics conoid (4.35) through P_0 only at P_0, is also an example of a space-like surface. It is simple to show that the condition of the coefficients v and $\boldsymbol{n}_0$ for (4.36) to be space-like is [47]

$$\frac{v^2}{|\boldsymbol{n}|^2} > a_0^2. \tag{4.37}$$

If we choose $|\boldsymbol{n}| = 1$, then v is the normal velocity of the moving plane (4.36) in $\boldsymbol{x}$-space. Thus (4.37) implies that a space-like plane (4.36) represents a locus in $\boldsymbol{x}$-space of points starting from $\boldsymbol{x}_0$ at t_0 and moving with a speed greater than the speed a_0. If we move along a curve on a space like surface R, then we move with a supersonic velocity (borrowing a word from gas dynamics).

Time-like direction and curve: Consider a straight line in space–time passing through a point P_0 $(\boldsymbol{x}_0, t_0)$. If the straight line lies in the interior of the characteristic conoid through the point P_0, then the direction of the straight line is said to be a **time-like direction** for the wave equation. A curve

$$\boldsymbol{x} = \boldsymbol{x}(\sigma) \ , \ t = t(\sigma) \tag{4.38}$$

in space–time is said to be a **time-like curve** if its tangent direction is always a time-like direction. This implies

$$\left(\frac{dt}{d\sigma}\right)^2 - \frac{1}{a_0^2}\left|\frac{d\boldsymbol{x}}{d\sigma}\right|^2 > 0. \tag{4.39}$$

If we move along a time like curve, we move with a subsonic velocity, for example moving along the t-axis means moving with zero velocity in physical space.

A generator of a characteristic conoid is neither space-like nor time-like and is called **null line**.

Bicharacteristics and Rays

Consider one parameter family of characteristic surfaces: $\Omega = \varphi(\boldsymbol{x}, t) = \alpha$, of the wave equation (4.32) so that the function φ satisfies the characteristic partial differential equation (4.33), i.e. $Q(\nabla\varphi, \varphi_t) = 0$. The characteristic curves of this nonlinear first-order equation in space–time are defined to be the **bicharacteristics** of the wave equation. They satisfy the Charpit's equations of $Q(\nabla\varphi, \varphi_t) = 0$, which are a system of ordinary differential equations for t, $\boldsymbol{x}$, φ_t and $\nabla\varphi = (\varphi_{x_1}, \varphi_{x_2}, \ldots, \phi_{x_d})$ with a constraint $Q(\nabla\varphi, \varphi_t) = 0$:

$$\frac{dt}{d\sigma} = \frac{1}{2}Q_{\varphi_t} = \varphi_t, \quad \frac{dx_\alpha}{d\sigma} = \frac{1}{2}Q_{\varphi_\alpha} = -a_0^2\varphi_{x_\alpha}, \tag{4.40}$$

$$\frac{d\varphi}{d\sigma} = \frac{1}{2}\varphi_t Q_{\varphi_t} + \frac{1}{2}\varphi_{x_\alpha}Q_{\varphi_{x_\alpha}} = Q(\nabla\varphi, \varphi_t) = 0, \quad \frac{d\varphi_t}{d\sigma} = -\frac{1}{2}Q_t = 0, \quad \frac{d\varphi_{x_\alpha}}{d\sigma} = -\frac{1}{2}Q_{x_\alpha} = 0. \tag{4.41}$$

We choose a particular characteristic surface Ω corresponding to $\alpha = 0$. Successive positions Ω_t of a wavefront of (4.32) form a one-parameter family of surfaces in $(x_1, \ldots, x_d)$-space given by $\varphi(x_\alpha, t) = 0$ with t as the parameter.

A **ray** associated with a wavefront Ω_t is a curve in the $\boldsymbol{x}$-space traced by a moving point $\boldsymbol{x}(\sigma)$ according to the Eqs. (4.40) and (4.41) and starting from a point of a particular wavefront, say the wavefront Ω_0 at time $t = 0$. Thus, a ray is the projection of a bicharacteristic on $\boldsymbol{x}$-space. Similarly, the wavefront Ω_t is the projection on the $\boldsymbol{x}$-space of the section of the characteristic surface Ω by a $t =$ constant plane. Thus, the moving point on a ray remains on the wavefront Ω_t at all times t. The equations of a point $\boldsymbol{x}$ on a *ray* of the wave equation (4.32) and the unit normal $\boldsymbol{n}$ of the wavefront at that point, can be derived from (4.40) and (4.41) using $n_\alpha = \varphi_{x_\alpha}/|\text{ grad }\varphi|$ and $\varphi_t = -a_0|\text{ grad }\varphi|$ (for a forward facing wavefront):

$$\frac{dx_\alpha}{dt} = n_\alpha a_0, \quad \frac{dn_\alpha}{dt} = 0. \tag{4.42}$$

Therefore, the rays of the wave equation starting from a point $\boldsymbol{x}_0$ at time $t = t_0$ are straight lines normal to the successive positions of the wavefront:

$$\boldsymbol{x} = \boldsymbol{x}_0 + \boldsymbol{n}a_0(t - t_0). \tag{4.43}$$

Caustic Formation on Linear Wavefront

Singularities on wavefronts are very common physical phenomena. Consider, a wavefront Ω_0 at $t = 0$ having a concave part and governed by the the characteristic PDE (4.33) of wave equation (4.32), with $a_0 = 1$. In general, the converging rays from the points on the concave part of Ω_0 envelope a surface, called **caustic**, on which the successive positions of the wavefront have a cusp type of singularity. Usually, a caustic itself starts with a cusp, called **arête** (the point where the rays first meet), and its two branches (in two-space dimensions) bound a region in which the wavefront folds and crosses itself (see Fig. 9.1A in Chap. 9). An interesting example of a caustic appears during the propagation a 2-D wavefront, which is initially given by

$$\left.\begin{array}{ll} y^2 = 4x & , \quad 0 \le x \le 1 \\ y = \pm(x+1) & , \quad x > 1 \end{array}\right\}. \tag{4.44}$$

The central part of the initial wavefront is a parabola extended by its tangents beyond $x > 1$. It has a continuously turning tangent even at the points $B(1, \pm 2)$.

Details of the solution are given given in [42]. The Fig. 9.1A shows the successive positions of the wavefront Ω_t. The arête of this caustic is at (2, 0). The two branches of the caustic, starting at time $t = 2$, end at a finite distance from the arête at $t = 4\sqrt{2}$. The lower branch of the caustic is enveloped by the rays from the upper part of the initial wavefront and extends from (2, 0) to (5, −2). The wavefront reaches the arête at $t = 2$. *An interesting part of this example is the existence of a cusp type of singularity on the wavefront for $t > 4\sqrt{2}$ even though there is no caustic.* This singularity on the wavefront results from the discontinuity in the curvature of the initial wavefront at the point B.

4.3.2 Hyperbolic System in Multi-dimensions

In order to define a hyperbolic system, say of first-order PDEs, in $d+1$ independent variables, we need to look for the existence of time-like coordinates and space-like d-dimensional surfaces. A good classical discussion of this is available in [13] and also in [42, 47]. Once we are able to do this, we get a system of n first-order equations in the form (1.4), where it is assumed that $A \neq 0$ in a domain D of the **space–time**. Since we straight way go to a quasilinear system (1.4), the domain D and all results which we shall discuss are valid locally for a given solution $\boldsymbol{u}(\boldsymbol{x}, t)$ of the equation.

Definition 4.3.1 The system (1.4) is defined to be hyperbolic in D with t as time-like variable if, given an arbitrary unit vector $\boldsymbol{n}$, the characteristic equation

$$Q(\boldsymbol{x}, t, \boldsymbol{u}, \boldsymbol{n}) \equiv \det\left[n_\alpha B^{(\alpha)} - cA\right] = 0 \tag{4.45}$$

has n real roots (called eigenvalues) and eigenspace is complete at each point of D.

We denote the eigenvalues as

$$c_1, c_2, \ldots, c_n \tag{4.46}$$

and left and right eigenvectors by $\boldsymbol{\ell}^{(i)}$ and $\boldsymbol{r}^{(i)}$, which satisfy

$$\boldsymbol{\ell}^{(i)}\,(n_\alpha B^{(\alpha)}) = c_i \boldsymbol{\ell}^{(i)} A, \;\; (n_\alpha B^{(\alpha)})\boldsymbol{r}^{(i)} = c_i A \boldsymbol{r}^{(i)}. \tag{4.47}$$

Suppose an eigenvalue $c_i(\boldsymbol{x}, t, \boldsymbol{u}, \boldsymbol{n})$ is repeated p_i times in the set (4.46), completeness of eigenspace at each point of D implies that the number of linearly independent left eigenvectors (and hence also right eigenvectors) corresponding to c_i is p_i. Each of the left and right eigenvectors $\boldsymbol{\ell}^{(i)}$, $\boldsymbol{r}^{(i)}$ is unique except for a scalar multiplier. It also implies that we have a hyperbolic system (1.4) with **characteristics of uniform constant multiplicity**.

Definition 4.3.2 Any quantity associated with the ith eigenvalue c_i will be referred to as the quantity belonging to ***i*th characteristic field**.

Thus, we shall refer to $\boldsymbol{\ell}^{(i)}$ as the left eigenvalue of the ith characteristic field.

Polytropic gas: The most important example of a hyperbolic system, which we shall consider, is the system of **Euler's equations**, which governs the unsteady motion of a compressible medium, where it is assumed that irreversible processes like viscosity and heat conduction are absent [14]. Generally, one considers an **ideal gas**, in which the pressure p, density ρ and temperature T are related by the equation of state $p = R\rho T$, where R is the universal gas constant. In an ideal gas **internal energy** is a function of the temperature alone. In particular, we shall be concerned with a **polytropic gas** ([14], Sect. 3) for which the **specific internal energy** e is proportional to the temperature, i.e. $e = c_v T$, where the constant c_v is the specific heat at a constant volume. Another important thermodynamic variable is specific entropy, which for a polytropic gas is given by $\sigma = c_v log(\frac{p}{(\gamma-1)\rho^\gamma})$ + a constant. Euler's equations governing the motion of a polytropic gas in two and three space variables, represent an elegant and most important example of a hyperbolic system of quasilinear equations with a multiple eigenvalue. For all applications of KCL theory, we shall choose this system.

Example 4.3.3 ***Euler's equations of motion of a polytropic gas:*** *In three-dimensions, the equations are*

$$\rho_t + \langle \boldsymbol{q}, \nabla\rho \rangle + \rho\langle \nabla, \boldsymbol{q} \rangle = 0, \tag{4.48}$$

$$\boldsymbol{q}_t + \langle \boldsymbol{q}, \nabla \rangle \boldsymbol{q} + \frac{1}{\rho}\nabla p = 0, \tag{4.49}$$

$$p_t + \langle \boldsymbol{q}, \nabla \rangle p + \rho a^2 \langle \nabla, \boldsymbol{q} \rangle = 0, \tag{4.50}$$

where ρ is the mass density, $\boldsymbol{q} = (q_1, q_2, q_3)$ the fluid velocity, p the pressure, a is sound velocity in the medium given by

$$a^2 = \gamma p/\rho \tag{4.51}$$

and γ is the ratio of specific heats, assumed to be constant. It is a system of 5 first-order quasilinear equations. Taking $\boldsymbol{u} = (\rho, q_1, q_2, q_3, p)^T$, we find $A = I$ and the matrix

$$B^{(\alpha)} = \begin{bmatrix} q_\alpha & \rho\delta_{1\alpha} & \rho\delta_{2\alpha} & \rho\delta_{3\alpha} & 0 \\ 0 & q_\alpha & 0 & 0 & \rho^{-1}\delta_{1\alpha} \\ 0 & 0 & q_\alpha & 0 & \rho^{-1}\delta_{2\alpha} \\ 0 & 0 & 0 & q_\alpha & \rho^{-1}\delta_{3\alpha} \\ 0 & \rho a^2\delta_{1\alpha} & \rho a^2\delta_{2\alpha} & \rho a^2\delta_{3\alpha} & q_\alpha \end{bmatrix}. \tag{4.52}$$

The five eigenvalues are

$$c_1 = \langle \boldsymbol{n}, \boldsymbol{q} \rangle - a, \ c_2 = c_3 = c_4 = \langle \boldsymbol{n}, \boldsymbol{q} \rangle, \ c_5 = \langle \boldsymbol{n}, \boldsymbol{q} \rangle + a. \tag{4.53}$$

We can easily check that there are three linearly independent left (or right) eigenvectors corresponding to the triple eigenvalue $\langle \boldsymbol{n}, \boldsymbol{q} \rangle$ so that the system (4.48)–(4.50) is hyperbolic.

For later use, we give the system of conservation laws representing conservation of mass, the three components of momentum and energy of gas elements from which the Eqs. (4.48–4.50) have been derived. These are

$$\boldsymbol{H}_t + \langle \nabla, \boldsymbol{F} \rangle = 0 \tag{4.54}$$

$$\textit{where, } \boldsymbol{H} = \begin{bmatrix} \rho \\ \rho q_1 \\ \rho q_2 \\ \rho q_3 \\ \rho(e + \frac{1}{2}\boldsymbol{q}^2) \end{bmatrix}, \quad \boldsymbol{F} = \begin{bmatrix} \rho \boldsymbol{q} \\ \rho(q_1^2 + p/\rho,\ q_1 q_2,\ q_1 q_3) \\ \rho(q_2 q_1,\ q_2^2 + p/\rho,\ q_2 q_3) \\ \rho(q_3 q_1,\ q_3 q_2,\ q_3^2 + p/\rho) \\ \rho \boldsymbol{q}(e + \frac{p}{\rho} + \frac{1}{2}\boldsymbol{q}^2) \end{bmatrix} \tag{4.55}$$

The specific internal energy (denoted by e) and the relation between pressure, density and specific entropy (denoted by σ) for a **polytropic gas** *are given by*

$$e = \frac{p}{(\gamma - 1)\rho} \quad \text{and} \quad p = A(\sigma)\rho^{\gamma}, \quad \textit{with } A(\sigma) = (\gamma - 1)\exp\left(\frac{\sigma - \sigma_0}{c_v}\right), \tag{4.56}$$

where σ_0 is a constant.

4.3.3 Characteristic Surface, Bicharacteristics and Rays

Let $\Omega : \varphi(\boldsymbol{x}, t) = \alpha$, $\alpha =$ constant be a one parameter family of characteristic surfaces in a characteristic field of eigenvalue c of the system (1.4). Since $n_\alpha = \varphi_{x_\alpha}/|\nabla\varphi|$ and $c = -\varphi_t/|\nabla\varphi|$, characteristic equation (4.45) of the system (1.4) takes the form of PDE

$$Q(\boldsymbol{x}, t; \nabla\varphi, \varphi_t) \equiv \det(A\varphi_t + B^{(\alpha)}\varphi_{x_\alpha}) = 0, \tag{4.57}$$

where we have not shown the dependence of Q on the known solution $\boldsymbol{u}$.

When $|\boldsymbol{n}| = 1$ in (4.45), the eigenvalue c is the **normal velocity** or simply **velocity** of a wavefront $\Omega_t : \varphi(\boldsymbol{x}, t) = \alpha$, ($t$ and $\alpha =$ constant) in $\boldsymbol{x}$-space.

Equation (4.57) is a first-order nonlinear partial differential equation for the function φ. The characteristic curves of (4.57) are called **bicharacteristics** of (1.4). These are curves in space–time whose parametric representation is obtained after solving the Charpit's ordinary differential equations of (4.57):

$$\frac{dt}{d\sigma} = \frac{1}{2}Q_q, \quad \frac{dx_\alpha}{d\sigma} = \frac{1}{2}Q_{p_\alpha} \tag{4.58}$$

and

$$\frac{dq}{d\sigma} = -\frac{1}{2}Q_t, \quad \frac{dp_\alpha}{d\sigma} = -\frac{1}{2}Q_{x_\alpha} \tag{4.59}$$

where

$$q = \phi_t, \quad p_\alpha = \phi_{x_\alpha}, \quad \alpha = 1, 2, \ldots, d,$$

and imposing a condition $Q(\boldsymbol{p}, q, \boldsymbol{x}, t) = 0$. We shall show below that for a given solution $\boldsymbol{u}$, the set of all bicharacteristics curves form a $2d - 1$ parameter family of curves in space–time.

As in Sect. 4.3.1, we define **rays** as the projections of the bicharacteristics on the hyperplane $t = 0$. These are curves in $\boldsymbol{x}$-space. We quote here an important result from [13].

Lemma 4.3.4 Lemma on bicharacteristics direction[4]: *Consider an eigenvalue c (assumed to be simple here) of the first-order system (1.4) with $\boldsymbol{\ell}$ and $\boldsymbol{r}$ as left and right eigenvectors satisfying (4.47). Then components χ_α of the corresponding ray velocity $\boldsymbol{\chi}$ are*

$$\chi_\alpha = \frac{\boldsymbol{\ell} B^{(\alpha)} \boldsymbol{r}}{\boldsymbol{\ell} A \boldsymbol{r}}. \tag{4.60}$$

We prove below a theorem, which also gives a simpler proof of the 'lemma on bicharacteristic direction'.

Theorem 4.3.5 *For an eigenvalue c of the system (1.4), the velocity $\boldsymbol{\chi}$ given by (4.60) satisfies the condition (4.5) and hence qualifies for being a ray velocity.*

Proof We take the first relation in (4.47) (dropping the subscript i from c, $\boldsymbol{\ell}$ and $\boldsymbol{r}$), post-multiply by $\boldsymbol{r}$ and use $n_\alpha = \varphi_{x_\alpha}/|\nabla\varphi|$ and $c = -\varphi_t/|\nabla\varphi|$. This gives the relation satisfied by ϕ_t and ϕ_{x_α} in the form (another form of the characteristic PDE)

$$\tilde{Q}(\boldsymbol{x}, t, \nabla\varphi, \varphi_t) \equiv (\boldsymbol{\ell} A \boldsymbol{r})\varphi_t + (\boldsymbol{\ell} B^{(\alpha)} \boldsymbol{r})\varphi_{x_\alpha} = 0. \tag{4.61}$$

This is exactly of the eikonal equation (4.4) provided we take $\boldsymbol{\chi}$ as given in (4.60). Now we need to show that this $\boldsymbol{\chi}$ satisfies the consistency condition (4.5).

We note that $\boldsymbol{\ell}$ and $\boldsymbol{r}$ (and hence $\boldsymbol{\chi}$) depend on $\boldsymbol{n}$ but A and $B^{(\alpha)}$ do not. Hence,

$$\begin{aligned}
n_\gamma \frac{\partial \chi_\gamma}{\partial n_\alpha} &= n_\gamma \frac{\partial}{\partial n_\alpha}\left(\frac{\boldsymbol{\ell} B^{(\gamma)} \boldsymbol{r}}{\boldsymbol{\ell} A \boldsymbol{r}}\right) \\
&= \frac{1}{(\boldsymbol{\ell} A \boldsymbol{r})^2}\left[n_\gamma\left(\frac{\partial}{\partial n_\alpha}(\boldsymbol{\ell})\right)\left\{\left(B^{(\gamma)}\boldsymbol{r}\right)(\boldsymbol{\ell} A \boldsymbol{r}) - (A\boldsymbol{r})\left(\boldsymbol{\ell} B^{(\gamma)}\boldsymbol{r}\right)\right\}\right] \\
&\quad + \frac{n_\gamma}{(\boldsymbol{\ell} A \boldsymbol{r})^2}\left[\left\{\left(\boldsymbol{\ell} B^{(\gamma)}\right)(\boldsymbol{\ell} A \boldsymbol{r}) - (\boldsymbol{\ell} A)\left(\boldsymbol{\ell} B^{(\gamma)}\boldsymbol{r}\right)\right\}\left(\frac{\partial}{\partial n_\alpha}(\boldsymbol{r})\right)\right] \\
&= \frac{1}{(\boldsymbol{\ell} A \boldsymbol{r})}\left[\left(\frac{\partial}{\partial n_\alpha}(\boldsymbol{\ell})\right)\left(n_\gamma B^{(\gamma)} - cA\right)\boldsymbol{r} + \boldsymbol{\ell}\left(n_\gamma B^{(\gamma)} - cA\right)\left(\frac{\partial}{\partial n_\alpha}(\boldsymbol{r})\right)\right] = 0,
\end{aligned} \tag{4.62}$$

[4] This name 'lemma on bicharacteristic direction' comes from [13].

where we have used (4.47) in the last line. Replacing α by β, we also get $n_\gamma \frac{\partial}{\partial n_\beta}\chi_\gamma = 0$. Therefore, the condition (4.5) is satisfied and χ given by (4.60) qualifies to be a ray velocity. ∎

Note: We have also proved that for an eigenvalue c, when $\boldsymbol{\ell}$ and $\boldsymbol{r}$ are determined from (4.47), we can treat the PDE (4.61) as a characteristic PDE instead of (4.57).

Now we state a very important theorem (first stated in [39]):

Theorem 4.3.6 *If χ be a given by (4.60), then the ray equations are*

$$\frac{dx_\alpha}{dt} = \frac{\boldsymbol{\ell}B^{(\alpha)}\boldsymbol{r}}{\boldsymbol{\ell}A\boldsymbol{r}} = \chi_\alpha \tag{4.63}$$

$$\frac{dn_\alpha}{dt} = -\frac{\boldsymbol{\ell}}{\boldsymbol{\ell}A\boldsymbol{r}}\boldsymbol{\ell}\left\{n_\beta\left(n_\gamma\frac{\partial B^{(\gamma)}}{\partial\eta^\alpha_\beta} - c\frac{\partial A}{\partial\eta^\alpha_\beta}\right)\right\}\boldsymbol{r} = \psi_\alpha,\ say, \tag{4.64}$$

where $\frac{\partial}{\partial\eta^\alpha_\beta}$ is given by (4.7). Further, the system (1.4) implies a compatibility condition on a characteristic surface Ω in the form

$$\boldsymbol{\ell}A\frac{d\boldsymbol{u}}{dt} + \boldsymbol{\ell}(B^{(\alpha)} - \chi_\alpha A)\frac{\partial\boldsymbol{u}}{\partial x_\alpha} + \boldsymbol{\ell}C = 0, \tag{4.65}$$

where

$$\frac{d}{dt} = \frac{\partial}{\partial t} + \chi_\alpha\frac{\partial}{\partial x_\alpha} \quad and \quad \boldsymbol{\ell}(B^{(\alpha)} - \chi_\alpha A)\frac{\partial}{\partial x_\alpha} \tag{4.66}$$

are derivatives in directions tangential to Ω.

Proof The Eq. (4.63) has already been derived in Theorem 4.3.6.

To derive (4.64) we follow the procedure of differentiation in (4.62), but with respect to x_α instead of n_α. In the Eq. (4.7), the derivatives $\frac{\partial}{\partial\eta^\alpha_\beta}$ and hence $\frac{\partial}{\partial x_\alpha}$ operate on $\chi_\alpha = \frac{\boldsymbol{\ell}B^{(\alpha)}\boldsymbol{r}}{\boldsymbol{\ell}A\boldsymbol{r}}$, which is a very complicated expression in $\boldsymbol{x}$ and $\boldsymbol{n}$. However, the original Charpit equations, from which (4.7) has been derived, involves differentiation with respect to x_α keeping ϕ_t and $\nabla\phi$ fixed, i.e. $\boldsymbol{n}$ fixed. We use this information to simplify the right hand of (4.7). There are two terms which vanish due to (4.47), as it happened in (4.62). The third term which survives are those in which derivatives $\frac{\partial}{\partial x_\alpha}$ appear only on A and $B^{(\gamma)}$. Finally, we get

$$n_\gamma\frac{\partial\chi_\gamma}{\partial x_\alpha} = \frac{1}{\boldsymbol{\ell}A\boldsymbol{r}}\boldsymbol{\ell}\left[n_\gamma\frac{\partial B^{(\gamma)}}{\partial x_\alpha} - c\frac{\partial A}{\partial x_\alpha}\right]\boldsymbol{r}. \tag{4.67}$$

Similar result for $n_\gamma\frac{\partial\chi_\gamma}{\partial x_\beta}$. These two results lead to (4.64).

Now we derive the third part, i.e. (4.65) of the theorem. Pre-multiplying (1.4) by $\boldsymbol{\ell}$, we get $\boldsymbol{\ell}A\boldsymbol{u}_t + \boldsymbol{\ell}B^{(\alpha)}\boldsymbol{u}_{x_\alpha} + \boldsymbol{\ell}C = 0$. Using the expression (4.66) for $\frac{d}{dt}$, we rewrite it in the form

$$\boldsymbol{\ell} A \frac{d\boldsymbol{u}}{dt} + \boldsymbol{\ell}(B^{(\alpha)} - \chi_\alpha A)\frac{\partial \boldsymbol{u}}{\partial x_\alpha} + \boldsymbol{\ell} \boldsymbol{C} = 0. \tag{4.68}$$

This scalar equation is a linear combination of n equations in (1.4). Remark 4.3.7 below shows that it contains only tangential derivatives on the characteristic surface $\Omega : \phi = \mathit{constant}$ and hence it represents a compatibility condition on Ω.

We write (4.68) in the form

$$l_i A_{ij} \frac{du_j}{dt} + \tilde{\partial}_j u_j + l_i C_i = 0 \tag{4.69}$$

where

$$\tilde{\partial}_j = s_j^\alpha \frac{\partial}{\partial x_\alpha} \equiv l_i (B_{ij}^{(\alpha)} - \chi_\alpha A_{ij}) \frac{\partial}{\partial x_\alpha}. \tag{4.70}$$

Remark 4.3.7 The derivative $\tilde{\partial}_j$ on u_j in the second term of (4.69) is a special tangential derivative on the characteristic surface Ω, it is a tangential derivative also on the *wavefronts* Ω_t. This follows from

$$n_\alpha s_j^\alpha = l_i A_{ij}(c - n_\alpha \chi_\alpha) = 0, \quad \text{for each } j, \tag{4.71}$$

since $c = n_\alpha \chi_\alpha$.

The form (4.69) of the compatibility condition, containing only tangential derivatives on Ω, has a very special feature. The derivative $\frac{d}{dt}$ in the bicharacteristic direction is the only derivative in this equation which contains $\frac{\partial}{\partial t}$. The other n tangential derivatives $\tilde{\partial}_j$ $(j = 1, 2, \ldots, d)$ contain only spatial derivatives and can be expressed in terms of any $d - 1$ of the d tangential derivatives L_α, defined in terms of $\frac{\partial}{\partial \eta_\beta^\alpha}$ in (4.7) by

$$L_\alpha = n_\beta \frac{\partial}{\partial \eta_\beta^\alpha}, \qquad \alpha = 1, 2, \ldots, d. \tag{4.72}$$

The operator L_α can also be written in the form

$$L_\alpha = n_\beta \left(n_\beta \frac{\partial}{\partial x_\alpha} - n_\alpha \frac{\partial}{\partial x_\beta} \right) = \frac{\partial}{\partial x_\alpha} - n_\alpha \left(n_\beta \frac{\partial}{\partial x_\beta} \right), \ \textit{i.e.}\ \boldsymbol{L} = \nabla - \boldsymbol{n}\langle \boldsymbol{n}, \nabla \rangle. \tag{4.73}$$

We note that the operator $\boldsymbol{L}$ is obtained by subtracting from ∇ its component in the normal direction of Ω_t. Hence, $\boldsymbol{L}$ is linear combination of tangential derivatives on Ω_t.

Example 4.3.8 ***A compatibility condition for the Euler's equations of a polytropic gas:*** *We consider here the eigenvalue $c_5 = \langle \boldsymbol{n}, \boldsymbol{q} \rangle + a$ of the system of Eqs. (4.48)–(4.50). Left and right eigenvectors $\boldsymbol{\ell}$ and $\boldsymbol{r}$ corresponding to c_5 can be chosen to be*

$$\boldsymbol{\ell} = (0, n_1, n_2, n_3, \frac{1}{\rho a}), \quad \boldsymbol{r} = (\rho/a, n_1, n_2, n_3, \rho a). \tag{4.74}$$

The characteristic partial differential equation (4.4) corresponding to this eigenvalue is

$$\tilde{Q} \equiv \phi_t + \langle \boldsymbol{q}, \nabla\phi \rangle + a|\nabla\phi| = 0. \tag{4.75}$$

The ray equations (4.63) and (4.64) become

$$\frac{d\boldsymbol{x}}{dt} = \boldsymbol{q} + \boldsymbol{n}a \tag{4.76}$$

and

$$\frac{d\boldsymbol{n}}{dt} = -\boldsymbol{L}a - n_\beta \boldsymbol{L}q_\beta. \tag{4.77}$$

Multiplying the equations in (4.48)–(4.50) by components of $\boldsymbol{\ell}$ and adding the results, we derive the compatibility condition on the characteristic surface as

$$a\frac{d\rho}{dt} + \rho\langle \boldsymbol{n}, \frac{d\boldsymbol{q}}{dt}\rangle + \rho a\langle \boldsymbol{L}, \boldsymbol{q}\rangle = 0, \tag{4.78}$$

where $\frac{d}{dt} = \frac{\partial}{\partial t} + \langle \boldsymbol{q} + a\boldsymbol{n}, \nabla\rangle$. This is the form of the compatibility condition (4.69) for the Euler's equations (4.48)–(4.50) for the characteristic velocity c_5.

4.4 Shocks, Jump Relations and Shock Manifold PDE

To deal with discontinuous solutions of a system of equations, we take up a system of conservation laws (1.8). We begin with its particular case in one space dimension, i.e. $d = 1$.

4.4.1 One-Dimensinal System of Conservation Laws

Consider

$$\boldsymbol{H}_t(\boldsymbol{u}) + \boldsymbol{F}_x(\boldsymbol{u}) = 0. \tag{4.79}$$

As in the case of a single conservation law (3.3), by (4.79) we mean an integral formulation like (3.1) or (3.2) and we can define a **weak solution** of (4.79) as we did for a single conservation law in Sect. 3.1.

Jump relations or RH conditions across a curve of discontinuity $\Omega : x = X(t)$ in a weak solution $\boldsymbol{u}$ of (4.79) can be derived exactly as in the Sect. 3.1. With the notations used there, we write the jump relations across a discontinuity for the system in the form

$$\dot{X}(t)[\boldsymbol{H}] = [\boldsymbol{F}] \quad or \quad S[\boldsymbol{H}] = [\boldsymbol{F}], \quad where\ S := \dot{X}(t). \tag{4.80}$$

This system of n equations for S gives n values of S as functions of $2n$ quantities $\boldsymbol{u}_\ell$ and $\boldsymbol{u}_r$. Thus we get one S_i for each of the ith characteristic field. Further, given the state $\boldsymbol{u}_r$, we can solve from these n equations $n+1$ quantities $u_{1\ell}, u_{2\ell}, \ldots, u_{n\ell}$ and S in terms of one of these and $\boldsymbol{u}_r$ (see such a result discussed for the Euler equations after (4.109)).

Shock and entropy condition: For a smooth solution $\boldsymbol{u}$, (4.79) is equivalent to the system (4.18). We assume that the eigenvalues $c_1, c_2, \ldots, c_n$ satisfy strict inequality

$$c_1(\boldsymbol{u}) < c_2(\boldsymbol{u}) < c_3(\boldsymbol{u}) < \cdots < c_n(\boldsymbol{u}), \quad \forall \boldsymbol{u} \in \mathbb{R}^n. \tag{4.81}$$

For a scalar conservation law, we have discussed in some detail in Sect. 3.2 that for a unique weak solution, the velocity $S = \dot{X}(t)$ of an admissible discontinuity should satisfy (3.15). A similar condition for the system (4.79) have been deduced on two considerations. First, the weak solution should be stable — we quote from [16] 'A manifestation of stability would be that smooth waves of small amplitude colliding with the shock are absorbed, transmitted and/or reflected as waves with small amplitude, without affecting the integrity of the shock itself, by changing substantially its strength or its speed of propagation' (originally worked out by Gel'fand [19], see also [47]). Second, the discontinuity works as an internal boundary and the conditions on the admissible boundary should be such that the solution should be uniquely determined ([51], Sect. D, Chap. 15). We sum up the essence of the results which 'admissible discontinuity' requires:
Case 1:

$$c_{k-1}(\boldsymbol{u}_\ell) < S < c_k(\boldsymbol{u}_\ell), \quad c_k(\boldsymbol{u}_r) < S < c_{k+1}(\boldsymbol{u}_r) \tag{4.82}$$

when the kth characteristic field is genuinely nonlinear (this result is to be carefully modified for $k=1$ and $k=n$, see (4.96)) and
Case 2:

$$c_k(\boldsymbol{u}_\ell) = S = c_k(\boldsymbol{u}_r) \tag{4.83}$$

when the kth characteristic field is linearly degenerate.
Note that (4.82) contains an inequality

$$c_k(\boldsymbol{u}_r) < S < c_k(\boldsymbol{u}_\ell) \tag{4.84}$$

which has a physical interpretation 'kth shock moves in kth characteristic field with supersonic velocity with respect to the state ahead and with subsonic velocity with respect to the state behind it'(see the example below).

Definition 4.4.1 The conditions (4.82) and (4.83) are called **Lax entropy conditions**.

Definition 4.4.2 A discontinuity satisfying the entropy condition (4.82) is called a **shock of kth characteristic family** and that satisfying the condition (4.83) is called a **contact discontinuity of kth characteristic family**

We state the condition (4.84) in the form of a theorem:

Theorem 4.4.3 *A constant state $\boldsymbol{u}_\ell$ on left can be connected to another constant state $\boldsymbol{u}_r$ on right by a shock of kth family only if $\delta = c_k(\boldsymbol{u}_r) - c_k(\boldsymbol{u}_\ell)$ is negative.*

Results in Theorems 4.2.12 and 4.4.3 helps us to solve the Riemann problem for a system of conservation laws (4.79) (for details see [20, 51] and Sect. C, Chap. 15 of [56]. We shall solve a Riemann problem for a 2-D WNLRT in Sect. 8.6.

In what follows we shall briefly summarize some results for one of the most important example of a physically realistic system of conservation laws. This will beautifully clarify all concepts we have discussed in this section. The modern mathematical theory of shock waves grew out of need to study this system, which has still many unsolved problems.

Euler equations of a compressible fluid: Consider the system of conservation laws governing the 1-D motion a polytropic gas (a particular case of (4.54)):

$$\begin{bmatrix} \rho \\ \rho q \\ \rho(e + \frac{1}{2}q^2) \end{bmatrix}_t + \begin{bmatrix} \rho q \\ \rho(q^2 + \frac{p}{\rho}) \\ \rho q(e + \frac{p}{\rho} + \frac{1}{2}q^2) \end{bmatrix}_x = \mathbf{0}. \tag{4.85}$$

Differential form of this system is a particular case of (4.48)–(4.50) with (4.56). This 3×3 system is hyperbolic with three distinct eigenvalues

$$c_1 = q - a \ < \ c_2 = q \ < \ c_3 = q + a, \tag{4.86}$$

where we can show that the first and third characteristic fields are genuinely nonlinear and the second characteristic field is linearly degenerate.

Theory of shock waves started with the study of the gas dynamic equations in 1848 but a shock was well understood only in 1910 (see [14], Sect. 51). Though the consequences of jump relations across a shock for the system (4.85) have been discussed in great detail in many books, for example [14], still it is not easy to present essence of these results briefly. Let us derive and review those, which are of important for our discussion in this book.

Let V_ℓ and V_r be velocities of the fluid on the left and on the right of the shock relative to the shock, which moves with velocity $S := \dot{X}(t)$, i.e.

$$V_\ell = q_\ell - S \ \text{ and } \ V_r = q_r - S. \tag{4.87}$$

The jump relations for the system (4.85) are

$$\rho_\ell V_\ell = \rho_r V_r = \ m, \ \text{ say,} \tag{4.88}$$

$$mq_\ell + p_\ell = mq_r + +p_r, \tag{4.89}$$

$$m(\frac{1}{2}q_\ell^2 + e_\ell) + q_\ell p_\ell = m(\frac{1}{2}q_r^2 + e_r) + q_r p_r. \tag{4.90}$$

Subtracting mS from the both sides of (4.89), we get

$$mV_\ell + p_\ell = mV_r + p_r. \tag{4.91}$$

Using $q_\ell = V_\ell + S$ and $q_r = V_r + S$ in (4.90) and removing $\frac{1}{2}mS^2$ from both sides, we get

$$m(\frac{1}{2}V_\ell^2 + e_\ell) + mV_\ell S + V_\ell p_\ell + Sp_\ell = m(\frac{1}{2}V_r^2 + e_r) + mV_r S + V_r p_r + Sp_r.$$

Substitute $V_\ell p_\ell = mp_\ell/\rho_\ell$ and $V_r p_r = mp_r/\rho_r$ in this equation and note that the terms containing S cancel due to (4.89). Hence we get

$$m(\frac{1}{2}V_\ell^2 + e_\ell + p_\ell/\rho_\ell) = m(\frac{1}{2}V_r^2 + e_r + p_r/\rho_r). \tag{4.92}$$

There are two cases:
Case 1: when $m = 0$, from (4.88) it follows that the velocity S of discontinuity equals the fluid velocity on both sides, i.e. $q_\ell = S = q_r$. The discontinuity appears in the linearly degenerate characteristic field of the eigenvalue c_2 and is a **contact discontinuity** satisfying the condition (4.83) ([14], Sect. 56).
Case 2: when $m \neq 0$, from (4.92) we get a relation exactly in the form of **Bernoulli's law**

$$\frac{1}{2}V_\ell^2 + e_\ell + p_\ell/\rho_\ell = \frac{1}{2}V_r^2 + e_r + p_r/\rho_r = \frac{1}{2}\hat{q}^2, \quad say, \tag{4.93}$$

where $\hat{q}$ is the common limit speed of the fluid ([14], Sects. 14, 15 and 55), attained when the density $\rho \to 0$ on either side.

Mechanical jump relations: These are relations obtained from conservation of mass and momentum without using the conservation of energy and are valid for any compressible medium, not necessarily a polytropic gas. The relation (4.93) is also such a relation. Using (4.88) in (4.91), we get

$$m^2 = \rho_\ell \rho_r \frac{p_\ell - p_r}{\rho_\ell - \rho_r}. \tag{4.94}$$

Again using (4.88), we get

$$V_\ell V_r = \frac{p_\ell - p_r}{\rho_\ell - \rho_r}. \tag{4.95}$$

Consider now a discontinuity, which is a shock in the characteristic field $c_3 = q + a$, i.e. a **forward facing shock** moving into the state (ρ_r, q_r, p_r). As mentioned just above the Lax entropy condition (4.82) that it is to be carefully modified for $k = n, n = 3$ here. The Lax entropy condition for the shock in the 3rd characteristic family is (see (4.84)) simply

$$c_3(\boldsymbol{u}_r) = q_r + a_r < S < q_\ell + a_\ell = c_3(\boldsymbol{u}_\ell). \tag{4.96}$$

For a forward facing shock $-V_r = S - q_r = A$, say; then $A > 0$ and the relative velocities V_ℓ and V_r are negative. This implies that the mass flux m given by (4.88) across the shock is negative. Using $V_\ell = \rho_r/\rho_\ell V_r$ from (4.88), we get an expression of the shock velocity $A = -V_r$ relative to the fluid velocity ahead of the shock from

$$V_r^2 = \frac{\rho_\ell}{\rho_r}\frac{p_\ell - p_r}{\rho_\ell - \rho_r} = A^2, \quad A = S - q_r > 0. \tag{4.97}$$

Shock manifold PDE: A, given by (4.97), is the velocity of a forward facing shock relative to the fluid velocity ahead of the shock and hence if the equation of the shock be represented by $\mathcal{S}(x,t) = 0$, then $-\mathcal{S}_t/\mathcal{S}_x - q_r = A$ and hence $\mathcal{S}$ satisfies the the relation

$$\mathcal{S}_t + q_r\mathcal{S}_x + A\mathcal{S}_x = 0 \tag{4.98}$$

on $\mathcal{S}(x,t) = 0$. q_r, p_r and q_r are defined only ahead of the shock, i.e. for $x > \mathcal{S}$. Similarly, q_ℓ, p_ℓ and q_ℓ are defined only for $x < \mathcal{S}$. Assuming all these functions to be smooth in their respective domains, we extend them on the other side of $\mathcal{S}(x,t) = 0$ as smooth functions (see Maslov [31] and Prasad [37]). Smooth embedding implies that even if the extensions are not unique, their values and the values of their derivatives are unique on the shock $\mathcal{S}(x,t) = 0$. Now q_r and A (given by (4.96)) are defined as smooth functions in a neighbourhood of the curve $\mathcal{S}(x,t) = 0$ in (x,t)-plane and (4.98) becomes a PDE to determine the function $\mathcal{S}(x,t)$.

Definition 4.4.4 We call (4.98) a **shock manifold PDE**.

Some results for a polytropic gas: In the discussion above, we did not assume that the gas is a polytropic gas (see (4.56)). We note that $1 < \gamma < \frac{5}{3}$ (see Sect. 3 [14]). For a polytropic gas, gas pressure and density on the two sides of the shock are related by (see [14], equation number (67.01))

$$\frac{p_\ell}{p_r} = \frac{(\gamma+1)\rho_\ell - (\gamma-1)\rho_r}{(\gamma+1)\rho_r - (\gamma-1)\rho_\ell} \quad or \quad \frac{\rho_\ell}{\rho_r} = \frac{(\gamma+1)p_\ell + (\gamma-1)p_r}{(\gamma+1)p_r + (\gamma-1)p_\ell} \tag{4.99}$$

and the Bernoulli's law (4.93) becomes (for derivation see [14], Sect. 66)

$$\frac{\gamma-1}{\gamma+1}\hat{q}^2 = \frac{2}{\gamma+1}a_\ell^2 + \frac{\gamma-1}{\gamma+1}V_\ell^2 = \frac{2}{\gamma+1}a_r^2 + \frac{\gamma-1}{\gamma+1}V_r^2 = \frac{p_\ell - p_r}{\rho_\ell - \rho_r} = a_*^2 \tag{4.100}$$

where a_* is the common critical speed on the two sides of the discontinuity. The critical speed a_* is the speed in a steady flow when the sound velocity a equals the relative velocity $|V|$ (see [14], sections 15 and 66). The relation (4.95) i.e., $V_\ell V_r = \frac{p_\ell - p_r}{\rho_\ell - \rho_r} = a_*^2$ is known as **Prandtl's relation**. (4.100) gives the relation between $\hat{q}$ and a_* for a polytropic gas (see [14], Section 14). From (4.97) we also note that $A^2 = \frac{\rho_\ell}{\rho_r} a_*^2$.

Let us define the shock Mach number as $M_r = \frac{|V_r|}{a_r}$ and quote an important relation (67.07) of [14] using the notation $\beta^2 = \frac{\gamma-1}{\gamma+1} < 1$ (for which the notation is μ^2 in [14]):

$$M_r = \frac{|V_r|}{a_r}, \quad \frac{p_\ell}{p_r} = (1 + \frac{\gamma - 1}{\gamma + 1})M_r^2 - \frac{\gamma - 1}{\gamma + 1} = (1 + \beta^2)M_r^2 - \beta^2. \tag{4.101}$$

From the part $q_r + a_r < S$ of the entropy condition (4.96) it follows that $M_r > 1$ and we get

$$1 < \frac{p_\ell}{p_r} < \infty. \tag{4.102}$$

Substituting (4.101) in the inverse of the expression for $\frac{\rho_\ell}{\rho_r}$ in (4.99), we can derive

$$\frac{\rho_r}{\rho_\ell} = \beta^2 + \frac{1 - \beta^2}{M_r^2}. \tag{4.103}$$

Lax entropy condition, which implies $M_r > 1$, gives

$$\frac{\rho_r}{\rho_\ell} < 1. \tag{4.104}$$

We have quoted expressions for the temperature T and entropy σ in the paragraph just above the equation (4.48). We now derive expressions for the temperature ratio $\frac{T_\ell}{T_r}$ and entropy jump $[\sigma] = \sigma_\ell - \sigma_r$ for a shock. Using (4.101) and (4.103) we can get

$$\frac{T_\ell}{T_r} = \frac{p_\ell \rho_r}{p_r \rho_\ell} = 1 - 2\beta^4 + \beta^2(1 + \beta^2)M_r^2 - \frac{\beta^2(1 - \beta^2)}{M_r^2} \tag{4.105}$$

and

$$\sigma_\ell - \sigma_r = c_v log\{(\frac{p_\ell}{p_r})(\frac{\rho_r}{\rho_\ell})^\gamma\} = c_v log\left[\{(1 + \beta^2)M_r^2 - \beta^2\}\left(\beta^2 + \frac{1 - \beta^2}{M_r^2}\right)^\gamma\right]. \tag{4.106}$$

For $M_r = 1$, $\frac{T_\ell}{T_r} = 1$. We write (4.105) in the form

$$\frac{T_\ell}{T_r} = 1-2\beta^4+\beta^2(1+\beta^2)+\beta^2(1+\beta^2)(M_r^2-1)-\beta^2(1-\beta^2)+\beta^2(1-\beta^2)\left(1-\frac{1}{M_r^2}\right)$$

$$= 1+\beta^2(1+\beta^2)(M_r^2-1)+\beta^2(1-\beta^2)\left(1-\frac{1}{M_r^2}\right).$$

For $M_r > 1$, $M_r^2 - 1 > 0$ and $1 - \frac{1}{M_r^2} > 0$, hence

$$\frac{T_\ell}{T_r} > 1 \quad for \quad M_r > 1. \tag{4.107}$$

We note that $\sigma_\ell - \sigma_r = 0$ for $M_r = 1$. It should be possible to verify from the expression (4.106) that $\sigma_\ell > \sigma_r$ when $M_r > 1$, but we refer to an elegant proof of Weyl ([14], pp. 144–145) for a very general gas satisfying Assumptions (66.1)–(66.3) of [14]. These are also satisfied by a polytropic gas. We state in the form of a theorem the result III in the second paragraph on page 144 of [14] in terms of a Hugoniot curve[5] of Weyl's general gas:

Theorem 4.4.5 *Along the whole Hugoniot curve the entropy increases with decreasing specific volume.*

Decreasing specific volume means increasing density ρ. We collect all results (4.102), (4.104), (4.107) and the result of the Theorem 4.4.5 in form a final important theorem:

Theorem 4.4.6 *The Lax entropy condition (4.96) implies that the shock is a compressive discontinuity, across which the gas pressure p, gas density ρ, temperature T and entropy σ increase as the fluid elements cross the shock from right to left for a forward facing shock.*

Using $V_\ell = q_\ell - S$ and $V_r = q_r - S$ in the Prandtl's relation $V_\ell V_r = a_*^2$ and the Bernoulli's law (4.100), we can derive a relation ([14], equation (67.10))

$$\frac{q_\ell - q_r}{a_r} = \frac{2}{\gamma+1}\left(\frac{S-q_r}{a_r} - \frac{a_r}{S-q_r}\right) = \frac{2}{\gamma+1}\left(\frac{A}{a_r} - \frac{a_r}{A}\right), \tag{4.108}$$

which, for a forward facing shock, gives the fluid velocity q_ℓ behind the shock in terms of the shock velocity S, the fluid velocity q_r and local sound velocity a_r ahead of the shock.

We may define a shock strength in many ways. For a forward facing shock, the state (ρ_r, q_r, p_r) on the right side of the shock is known. We define a shock strength μ by

[5] Hugoniot curve is the graph of p versus specific volume $\frac{1}{\rho}$ representing the jump relation between p and ρ across a shock.

$$\mu = \frac{\rho_\ell - \rho_r}{\rho_r}. \tag{4.109}$$

Equation (4.99) shows that we can express p_ℓ in terms of μ. Then we can use (4.97) to express the shock velocity A, relative to the fluid velocity in the state ahead, in terms of μ and finally, we can use (4.108) to express q_ℓ in terms of μ. Thus, the state just behind the shock can be determined if the state ahead of the shock and shock strength are known.

4.4.2 Multi-dimensional Systems of Conservation Laws

Derivation for jump relations for a curved shock governed by (1.8) are rarely described in standard books. Two different approaches to derive these are available, one in [21] (Sect. 2.2) and another in [42] (Sect. 3.5). Let the equation of the shock surface S_t at a time t in $\boldsymbol{x}$-space be given by $\mathcal{S}(\boldsymbol{x}, t) = 0$ with t = constant, then for the system (1.8), the jump relation or RH condition is

$$\mathcal{S}_t[\boldsymbol{H}] + \mathcal{S}_{x_\alpha}[\boldsymbol{F}^\alpha] = \boldsymbol{0} \tag{4.110}$$

or, in terms of the velocity $S = -\mathcal{S}_t/|\nabla_{\boldsymbol{x}}\mathcal{S}|$ of S_t and its unit normal $\boldsymbol{N} = \nabla_{\boldsymbol{x}}\mathcal{S}/|\nabla_{\boldsymbol{x}}\mathcal{S}|$ we get

$$-S[\boldsymbol{H}] + N_\alpha[\boldsymbol{F}^\alpha] = \boldsymbol{0}. \tag{4.111}$$

The jump relation (4.111) is exactly the same as the jump relation (4.80) of the 1-D conservation law (4.79) provided we replace the flux function $\boldsymbol{F}$ in (4.80) by the normal component $N_\alpha \boldsymbol{F}^\alpha$ of the flux vector $\tilde{\mathbf{F}}$ in (1.7). Note that $[\boldsymbol{N}] = 0$. After the presentation of these results for a general system (1.8), we need to take a particular example to discuss the kinematics of a multi-dimensional shock front. We choose the conservation laws (4.55) from which the Euler equations have been derived.

Shock rays of in a polytropic gas: Kinematics of a shock front requires derivation of the eikonal equation or shock manifold PDE governing the evolution of a shock front.

We first ask a more basic question. What are shock rays or what is a shock ray velocity? One obvious answer to this question for a shock front in a gas is

$$\chi = \mathbf{q}_r + \boldsymbol{N}A, \tag{4.112}$$

where $\mathbf{q}_r$ is the fluid velocity ahead of the shock, $\boldsymbol{N}$ the unit normal to the shock front and A is normal speed of the shock relative to the gas ahead of it. The expression (4.112) of a shock ray velocity, associated with the shock surface Ω in space–time represented by $\mathcal{S}(\boldsymbol{x}, t) = 0$, requires that $\mathcal{S}(\boldsymbol{x}, t)$ satisfies a **shock manifold partial differential equation** (SME) (after extending the functions as explained below (4.98))

$$\tilde{\mathcal{Q}}_{sh} \equiv \mathcal{S}_t + \langle \mathbf{q}_r, \nabla \rangle \mathcal{S} + A|\nabla \mathcal{S}|^2 = 0. \tag{4.113}$$

When we try to derive the equation (4.113) from the jump relations or RH conditions of the conservation laws (4.55), we run into difficulty. There are many other relations, which can be derived from (4.111) and each one them would claim to an SME. For example, the Prandtl relation (4.95) for a curved shock, when expressed in terms of derivatives of $\mathcal{S}$, becomes

$$\{\mathcal{S}_t + \langle \mathbf{q}_\ell, \nabla \rangle \mathcal{S}\} \{\mathcal{S}_t + \langle \mathbf{q}_r, \nabla \rangle \mathcal{S}\} - a_*^2 |\nabla \mathcal{S}|^2 = 0, \ \ a_*^2 = \frac{p_\ell - p_r}{\rho_\ell - \rho_r}. \tag{4.114}$$

The ray equations derived from the eikonal equation (4.114) are different from those derived from (4.113). The question arises: 'are the shock ray velocities (or more precisely the complete set of shock ray equations) obtained from different SMEs the same?'

The concept of SME and their equivalence in the above sense was first discussed by Prasad (1982) in [37]. Making further use of the RH conditions, it was shown that the shock ray equations given by the two SMEs (4.113) and (4.114) are equivalent. This result was generalized by Roy and Ravindran (1988) for almost all SMEs.

The eikonal equation $\tilde{Q} = 0$, i.e. (4.75) for a forward facing wavefront and that $\tilde{\mathcal{Q}}_{sh} = 0$, i.e. (4.113) for a forward facing shock are exactly the same except that $\boldsymbol{q}$ and a in the first are replaced by $\boldsymbol{q}_r$ and A in the second. Therefore, the shock ray equations are simple generalizations of the Eqs. (4.76) and (4.77):

$$\frac{d\boldsymbol{x}}{dt} = \boldsymbol{q}_r + \boldsymbol{N} A \tag{4.115}$$

and

$$\frac{d\boldsymbol{N}}{dt} = -\boldsymbol{L} A - N_\beta \boldsymbol{L} q_{r\beta}, \tag{4.116}$$

where $\boldsymbol{L} = \nabla - \boldsymbol{N}\langle \boldsymbol{N}, \nabla \rangle$.

Chapter 5
Equations of Nonlinear Wavefront and Shock Front

The aim of this research note is to present kinematical conservation laws and their applications. However, KCL forms an under-determined system. In order to get a determined system, we need to take a system where KCL can be applied. For this, we choose two challenging problems: propagation of a weak nonlinear wavefront[1] and that of a shock front in a polytropic gas in multi-dimensions. We ask the reader to the recollect the distinction between a nonlinear wavefront and a shock front in Sect. 3.5.

In a problem involving propagation a wavefront, we shall like to use the ray equations (4.63) and (4.64) and the transport equation (4.65) along with a bicharacteristic curve or ray. For a linear hyperbolic system, the ray equations (4.63) and (4.64) decouple from (4.65) and hence can be solved to give rays. For a quasilinear hyperbolic system (1.4), the matrices A and $B^{(\alpha)}$ depend on $\boldsymbol{u}$ and hence, the terms on the right sides of (4.63) and (4.64), when evaluated, would contain $\boldsymbol{u}$ and $\frac{\partial \boldsymbol{u}}{\partial \eta^{\alpha}_{\beta}}$. In this case, the coupled system of $2d$ equations[2] (4.63)–(4.65) in $2d + n - 1$ independent quantities in $\boldsymbol{x}, \boldsymbol{n}, \boldsymbol{u}$ is an under-determined unless $n = 1$. Therefore, this system is of limited use for $n > 1$. In order to get a determined system for the propagation of a nonlinear wavefront Ω_t, we need to recognize the wavefront by making a high-frequency approximation (Sects. 4.1 and 5.1 below). In this approximation, we can express the n components of $\boldsymbol{u}$ in terms of $\boldsymbol{x}, \boldsymbol{n}$ and a **wave amplitude** w and we get a determined system of equations governing $2d$ independent quantities in $\boldsymbol{x}, \boldsymbol{n}, w$ appearing in evolution equations of a nonlinear wavefront. This system turns out to be simple and in elegant form for a small amplitude wave (see [42], Chap. 4). We have called this system for a small amplitude as **weakly nonlinear ray theory** or simply **WNLRT**.

---

[1]Generally, in any book on wave propagation we use an intuitive idea of a wave and wavefront. A mathematical definition is available in Sect. 3.2.1 of [42], according to which a wave is an approximate concept in high-frequency approximation—see the next section. This precise definition was first given in 1977 [36].

[2]Note that only $d-1$ components of unit normal $\boldsymbol{n}$ are independent.

P. Prasad, *Propagation of Multidimensional Nonlinear Waves and Kinematical Conservation Laws*, Infosys Science Foundation Series,
https://doi.org/10.1007/978-981-10-7581-0_5

In order to find the successive positions of a shock front, we shall like to use the shock ray equations (4.115) and (4.116) derived from the SME (4.113). This SME contains the n components of the jump $\boldsymbol{u}_\ell - \boldsymbol{u}_r$. As in the case of a nonlinear wavefront, it appears that we shall have a highly under-determined system of equations. However, this is not the case,[3] as we can use (4.110) to determine the state on one side of the shock completely in terms of a shock strength and the state on the other side. We have seen this in the case of a polytropic gas (see a few lines just below equation (4.109)). Thus, when the state ahead of the shock is known the shock ray equations will have only one additional unknown, namely the shock strength. We need now sufficient[4] number of transport equations (along the shock ray), which along with the shock ray equations, form a closed system. Though gas dynamics shocks were well understood in early last century, the question of shock rays and transport equation of the shock amplitude along these rays were not considered till 1978, [22, 31]; and the derivation of shock rays from SME till 1982 [37]. We shall call a method of finding successive positions of a shock front using the shock rays and sufficient number of transport equations for the shock amplitude as **shock ray theory** or **SRT**.

5.1 Weakly Nonlinear Ray Theory - WNLRT

The high-frequency approximation mentioned above, when applied to a hyperbolic system of partial differential equations, gives a representation of the solution in terms of a wave amplitude w and a phase function $\varphi(\boldsymbol{x}, t)$ ([41] or [42] - Chap. 4). The high-frequency approximation further implies that

(i) the function φ satisfies an eikonal equation, with the help of which we can define rays,
(ii) all components of the state variables $\boldsymbol{u}$ of the system are expressed in terms of the an amplitude w and the unit normal $\boldsymbol{n}$ of the wavefront $\Omega_t : \varphi(\boldsymbol{x}, t) =$ constant, and in addition
(iii) when the expression for $\boldsymbol{u}$ in terms of w is substituted in (4.65) we get a transport equation for the amplitude w along the rays. This turns out to be an energy transport equation.

Thus, the high-frequency approximation reduces the problems of finding the successive positions of a wavefront and the amplitude distribution on it to an integration of a closed system of equations consisting of the ray equations and a transport equation. The ray equations and the transport equations are decoupled for a linear

[3] Unlike a wavefront, which is an approximate concept in high-frequency approximation, a shock in a solution of a system of conservation laws is an exact phenomenon and high-frequency approximation is exactly satisfied.

[4] We shall explain later the reason of using the word 'sufficient' for just one more unknown.

hyperbolic system and coupled for a quasilinear system. When the amplitude of the wave is a small perturbation over a known state $\boldsymbol{u}_0$, the relation between $\boldsymbol{u}-\boldsymbol{u}_0$ and w is linear. This leads to a simple weekly nonlinear ray theory (WNLRT) in which the equations can be integrated numerically, at least in theory. We review these important results below.

The high-frequency approximation was first applied in 1911 [13] to the wave equation

$$u_{tt} - a_0^2 \Delta u = 0, \quad \Delta = \sum_{\alpha=1}^{m} \frac{\partial^2}{\partial x_\alpha^2} \tag{5.1}$$

where a_0 is the constant sound velocity in the uniform medium. Small amplitude high-frequency approximation of a constant basic state $u = 0$ consists of assuming a solution of (5.1) in the form

$$u = \varepsilon u_1(\boldsymbol{x}, t, \theta) + \varepsilon^2 u_2(\boldsymbol{x}, t, \theta) + \cdots , \tag{5.2}$$

where the fast variable θ is related to the phase function φ by

$$\theta = \frac{1}{\varepsilon}\, \varphi(\boldsymbol{x}, t). \tag{5.3}$$

Substituting (5.2) in (5.1) and equating various powers of ε on both sides, we get a sequence of equations. The procedure of dealing with these equations is well known [43]. The leading order term gives the eikonal equation (4.33) from which we get the ray equations (4.41), i.e.

$$\frac{d\boldsymbol{x}}{dt} = \boldsymbol{n} a_0, \qquad \frac{d\boldsymbol{n}}{dt} = 0. \tag{5.4}$$

The next order term gives the transport equation for u_1, which leads to an equation relating u_1 and the ray tube area $\mathcal{A}$ (see [42], Sect. 2.2.3 or Sect. 7.1 later in this book)[5]

$$\frac{du_1}{dt} = a_0 \Omega u_1 = -\frac{1}{2\mathcal{A}} \frac{d\mathcal{A}}{dt} u_1. \tag{5.5}$$

Here, Ω is the mean curvature of the wavefront $\Omega_t : \varphi(\boldsymbol{x}, t) = 0$ and is given in terms of the unit normal $\boldsymbol{n}$ of Ω_t by

$$\Omega = -\frac{1}{2} div(\boldsymbol{n}). \tag{5.6}$$

The equation (5.5) gives $u_1 = u_{10}/\mathcal{A}^{1/2}$, u_{10} = constant. This is an example of a linear theory of wave propagation, where the high-frequency approximation

[5] In this section, for an element the arc length $d\ell$ of a linear ray, we have $d\ell = a_0 dt$. A proof of the result $\frac{1}{\mathcal{A}} \frac{d\mathcal{A}}{d\ell} = -\frac{1}{2}\Omega$ for a general ray system is given in Sect. 7.1.

(geometrical optics) gives a value of the leading term u_1 of the amplitude which tends to infinity as $\mathcal{A} \to 0$, i.e. as a focus or a caustic is approached. The small amplitude approximation implied in the expansion (5.2) breaks down wherever $\mathcal{A} \to 0$. In reality, the amplitude of the wave when evaluated more accurately does remain finite [11, 30]. We shall see later that the genuine nonlinearity present in a mode of propagation would make even the leading term u_1 of perturbation to remain finite and of the same order as the order of its initial value.

The basis of the weekly nonlinear ray theory applied to the quasilinear system (1.4) for a wave in a mode having genuine nonlinearity is to capture the wave amplitude $\tilde{w}$ ($= u_1$ of (5.2)) in the eikonal equation itself. One way to achieve this is to modify the expansion (5.2) by

$$\boldsymbol{u}(\boldsymbol{x}, t, \varepsilon) = \boldsymbol{u}_0 + \varepsilon \tilde{\boldsymbol{v}}\left(\boldsymbol{x}, t, \frac{\varphi(\boldsymbol{x}, t, \varepsilon)}{\varepsilon}, \varepsilon\right), \tag{5.7}$$

where

$$\tilde{\boldsymbol{v}}(\boldsymbol{x}, t, \theta, \varepsilon) = \tilde{\boldsymbol{v}}_0(\boldsymbol{x}, t, \theta, \varepsilon) + \varepsilon \tilde{\boldsymbol{v}}'(\boldsymbol{x}, t, \theta, \varepsilon), \quad \theta = \frac{\varphi}{\varepsilon} \tag{5.8}$$

for a perturbation of a basic solution $\boldsymbol{u} = \boldsymbol{u}_0$ of (1.4). The expansion (5.7) and (5.8) requires some explanation, in which we do not go (see [38, 41] and Sect. 4.4 of [42]). For our purpose in this book, we take a simpler system

$$A(\boldsymbol{u})\boldsymbol{u}_t + B^{(\alpha)}(\boldsymbol{u})\boldsymbol{u}_{x_\alpha} = 0, \tag{5.9}$$

choose $\boldsymbol{u}_0$ to be constant, substitute the above expansion in it and go through long mathematical steps carefully collecting terms up to order ε^2. This leads to a derivation of equations of WNLRT, where it turns out that leading term of perturbation $\tilde{\boldsymbol{v}}_0$ is given in terms of the wave amplitude $\tilde{w}$ (which is $\boldsymbol{O}(1)$) by

$$\tilde{\boldsymbol{v}}_0 = \boldsymbol{R}_0(\boldsymbol{u}_0)\tilde{w}.$$

A simple algorithm (see after (3.25) in [39] or [40]) to derive eikonal equation, ray and transport equations of WNLRT is to substitute $\boldsymbol{u}$ by $\boldsymbol{u}_0 + \varepsilon\tilde{\boldsymbol{v}}_0$ (4.61), (4.63)–(4.65) and retain terms only up to first order in ε.

WNLRT for a Polytropic Gas: We write the above WNLRT equations for the Euler equations (4.48)–(4.51) of a polytropic gas. We take the basic state to be a constant equilibrium state $\boldsymbol{u}_0 = (\rho_0, \boldsymbol{q} = \boldsymbol{0}, p_0)$ at rest, then the leading order perturbation $\varepsilon\tilde{\boldsymbol{v}}_0$, for the forward facing wavefront corresponding to the eigenvalue $c_5 = \langle \boldsymbol{n}, \boldsymbol{q}\rangle + a$, turns out to be

$$\varepsilon\tilde{v}_{10} = \rho - \rho_0 = \varepsilon\frac{\rho_0}{a_0}\tilde{w}, \quad \varepsilon(\tilde{v}_{20}, \tilde{v}_{30}, \tilde{v}_{40}) = \varepsilon\boldsymbol{n}\tilde{w}, \quad \varepsilon\tilde{v}_{50} = p - p_0 = \varepsilon\rho_0 a_0\tilde{w},$$
$$a = a_0\left(1 + \varepsilon\frac{\gamma - 1}{2}\tilde{w}\right), \tag{5.10}$$

where $a_0 = \sqrt{\gamma p_0/\rho_0}$ and $\tilde{w}$ has the dimension of velocity.[6] Retaining only the leading terms, the eikonal equation for φ (for the forward facing nonlinear wavefront) becomes

$$\varphi_t + \left(a_0 + \varepsilon\frac{\gamma+1}{2}\tilde{w}\right)|\nabla\varphi| = 0.$$

Remark 5.1.1 We now see that, we have captured the amplitude $\tilde{w}$ in the eikonal equation. This is in contrast to the nonlinear small amplitude perturbation of Whitham (see [56]), Choquet-Bruhat [12], Parker [34, 35] and Hunter and Keller [23] (see Sect. 4.2 in [42]), where the eikonal equation is $\varphi_t + a_0|\nabla\varphi| = 0$, which is a factor of (4.33).

The ray equations corresponding to this eikonal equation are given by[7]

$$\frac{d\boldsymbol{x}}{dt} = \left(a_0 + \varepsilon\frac{\gamma+1}{2}\tilde{w}\right)\boldsymbol{n}, \quad \frac{d\boldsymbol{n}}{dt} = -\varepsilon\frac{\gamma+1}{2}\boldsymbol{L}\,\tilde{w} \tag{5.11}$$

and the energy transport equation of the WNLRT, following the above algorithm, is

$$\frac{d\tilde{w}}{dt} \equiv \left\{\frac{\partial}{\partial t} + \left(a_0 + \varepsilon\frac{\gamma+1}{2}\tilde{w}\right)\langle\boldsymbol{n}, \nabla\rangle\right\}\tilde{w} = -\frac{1}{2}a_0\langle\nabla, \boldsymbol{n}\rangle\tilde{w} \equiv \Omega a_0\,\tilde{w} \tag{5.12}$$

where, as defined in (4.72),

$$\boldsymbol{L} = \nabla - \boldsymbol{n}\langle\boldsymbol{n}, \nabla\rangle \tag{5.13}$$

and Ω is now the mean curvature of the nonlinear wavefront Ω_t, **(the mean curvature here is not non-dimensionalized as in the Eq.** (5.53)).[8] The components L_α of the operator $\boldsymbol{L}$ represent tangential derivatives on the surface Ω_t. L_α can be expressed as a linear combination of the operators $\frac{\partial}{\partial\eta_\beta^\alpha}$ (see (4.72)).

5.2 Shock Ray Theory (SRT) - an Example in One Space Dimension

In the beginning of this chapter, we mentioned that the the eikonal equation for a shock, i.e. the SME is an exact equation and though not unique, gives a unique system of shock rays. The main problem now is to derive the compatibility conditions along

---

[6] $\tilde{w}$ in [32, 42] is non-dimensional.

[7] The term on the right-hand side of the second equation in (5.11) is missing in all references quoted in Remark 5.1.1. This term is necessary in order that the rays given by (5.11) satisfy Fermat's principle of stationary time of transit (see [42], Sect. 3.2.7).

[8] Note that the symbols Ω_t and Ω represent respectively a surface in $I\!R^d$ (here d =3) and mean curvature of Ω_t.

a shock ray. It is better to begin with the case of a scalar conservation law in one space dimension. Consider the simplest conservation law (1.2), i.e.

$$u_t + \left(\frac{1}{2}u^2\right)_x = 0. \tag{5.14}$$

Consider a shock along a curve $\Omega : x = X(t)$ in (x, t)-plane, then

$$\frac{dX(t)}{dt} = \frac{1}{2}(u_\ell + u_r) = shock\ velocity = S, \tag{5.15}$$

where u_ℓ and u_r are states on the left and right of the shock. The state on $x < X(t)$ satisfies $u_t + uu_x = 0$ which we write as,

$$u_t + \frac{1}{2}(u + u_r)\ u_x = -\frac{1}{2}(u - u_r)\,u_x. \tag{5.16}$$

Taking the limit $x \to X(t)-$, we get

$$\frac{du_\ell}{dt} = -\frac{1}{2}(u_\ell - u_r)\,(u_x)_\ell, \tag{5.17}$$

where

$$\frac{d}{dt} = \frac{\partial}{\partial t} + \frac{1}{2}(u_\ell + u_r)\frac{\partial}{\partial x} \equiv \frac{\partial}{\partial t} + S\frac{\partial}{\partial x} \tag{5.18}$$

represents time rate of change along the shock path given by (5.15).

Differentiating $u_t + uu_x = 0$ with respect to x we get,

$$(u_x)_t + u\,(u_x)_x = -u_x^2 \tag{5.19}$$

which we write as

$$(u_x)_t + \frac{1}{2}(u + u_r)\ (u_x)_x = -\frac{1}{2}(u - u_r)\ (u_x)_x - u_x^2 \tag{5.20}$$

and taking the limit as $x \to X(t)-$, we get

$$\frac{d(u_x)_\ell}{dt} = -\frac{1}{2}(u_\ell - u_r)\,(u_{xx})_\ell - ((u_x)_\ell)^2 \tag{5.21}$$

where, again $\frac{d}{dt} = \frac{\partial}{\partial t} + S\,\frac{\partial}{\partial x}$ represents the time rate of change as we move with the shock. Following this procedure, we derive an infinite system of compatibility conditions.

We define the limiting value of the ith derivative of u on the left of the shock divided by $i!$ by $v_i(t)$, i.e.

$$v_i(t) = \frac{1}{i} \lim_{x \to X(t)-} \frac{\partial^i u}{\partial x^i}, \quad i = 1, 2, 3, \cdots \tag{5.22}$$

and introduce $v_0(t) = u_\ell(t)$. The Eq. (5.15) and the infinite system of compatibility conditions along the shock path Ω are ([42], Sect. 7.1)

$$\frac{dX}{dt} = \frac{1}{2}\left(v_0 + u_r\right), \tag{5.23}$$

$$\frac{du_0}{dt} = -\frac{1}{2}\left(v_0 - u_r\right) v_1, \tag{5.24}$$

$$\frac{dv_i}{dt} = -\frac{i+1}{2}\left(v_0 - u_r\right) v_{i+1} - \frac{i+1}{2} \sum_{j=1}^{i} v_j v_{i-j+1}, \quad i = 1, 2, 3, \ldots \tag{5.25}$$

Given an initial value $u(x, 0) = u_0(x)$ for (5.14) with a single shock at X_0, we can derive initial values of $X(t)$, $v_0(t)$, $v_i(t)$, $i = 1, 2, 3, \ldots$ in the form

$$\begin{aligned} X(0) = X_0, \quad v_0(0) = u_0(0-) = u_{00}, \ say, \\ v_i(0) = v_{i0}, \ i = 1, 2, 3, \ldots \end{aligned} \tag{5.26}$$

However, we face new problems ([42], Sect. 7.2):

- The solution of an infinite system of equations (5.23)–(5.26) is more difficult than the Cauchy problem for the original conservation law (5.14).
- For an analytic initial value $u_0(x)$ for (5.14) in a neighbourhood of $x = X_0$, the analytic solution of the problem (5.23)–(5.26) is unique and give a function $u(x, t)$ by

$$u(x, t) = v_0(t) + \sum_{i=1}^{\infty} v_i(t)(x - X(t))^i \tag{5.27}$$

  which tends in a neighbourhood of $x = X(t)$ to the analytic solution of the initial value problem for (5.14) for small time.
- For a more general non-analytic initial data, the solution of initial value problem (5.23)–(5.26) is non-unique.

Ravindran and Prasad [49] proposed a new theory of shock dynamics (NTSD) by setting $v_{n+1} = 0$ in the nth equation in (2.25) so that the first $n + 1$ equations in (5.24) and (2.25) form a closed system. Then, the first $(n + 2)$ equations in (5.23)–(5.26) form a system of ordinary differential equations [49], which can be easily integrated numerically with the first $n + 2$ initial data in (5.26). The new theory of shock dynamic gives excellent results for the case when $u_0'(x) > 0$ behind the shock at $x = X_0$. There are open mathematical questions to be answered but this method has been very successful for many practical problems in multi-space dimensions.

5.3 SRT for a Shock in Multi-dimensions

Derivation of the compatibility conditions along a shock ray, even for Euler's equations (4.48)–(4.51), requires extremely complex calculations [50]. In fact, it is complex even in one-dimension ([42], Sect. 8.2). From these compatibility conditions, we can derive their simpler forms for a weak shock ([42], Sect. 9.5). An alternative derivation of the shock ray equations and compatibility conditions for a weak shock is very simple when we use the Eqs. (5.11) and (5.12) of the WNLRT ([32] or Sect. 10.1 in [42]). This method started with a derivation ray equations for a weak shock for Euler equations in [37]. An extension of the method is possible for a general system, which we state and prove in the form of a theorem in the next subsection.

5.3.1 A Theorem on Shock Ray Equations for a Weak Shock

We state a theorem[9] in the most general form for the hyperbolic system of conservation laws (1.8).

Theorem 5.3.1 *For a weak shock, the shock ray velocity is equal to the mean of the ray velocities of the nonlinear wavefronts just ahead and just behind the shock, provided we take the wavefronts ahead and behind to be instantaneously coincident with the shock surface. Similarly, the rate of turning of the shock front, i.e. dN/dt, is also equal to the mean of the rates of turning of such wavefronts just ahead and just behind the shock. These are approximate results correct up to first order in the shock strength.*

Remark 5.3.2 We note that the nonlinear waves of the kth family, on the two sides of the shock, need not be of small amplitude.

Proof We shall find an approximate form of the jump relations (4.110) of the system of conservation laws (1.8) assuming $|\boldsymbol{u}_\ell - \boldsymbol{u}_r| = \varepsilon$ small. We first note that

$$\boldsymbol{H}\left(\frac{1}{2}(\boldsymbol{u}_\ell + \boldsymbol{u}_r)\right) = \boldsymbol{H}(\boldsymbol{u}_\ell) - \frac{1}{2}\langle\langle \nabla_{\boldsymbol{u}}, \boldsymbol{H}\rangle(\boldsymbol{u}_\ell), (\boldsymbol{u}_\ell - \boldsymbol{u}_r)\rangle + O(\varepsilon^2) \tag{5.28}$$

[9] This theorem is reproduced from Sect. 3 of [45] with some changes. The author thanks Indian Academy Sciences for kindly giving permission for reproduction.

and

$$H\left(\frac{1}{2}(u_\ell + u_r)\right) = H(u_r) + \frac{1}{2}\langle\langle\nabla_u H\rangle(u_r), (u_\ell - u_r)\rangle + O(\varepsilon^2). \tag{5.29}$$

Subtracting and noting $A(u) = \langle\nabla_u, H\rangle$ *and* $B^{(\alpha)}(u) = \langle\nabla_u, F^{(\alpha)}\rangle$, we get

$$[H] = \frac{1}{2}\left(A(u_\ell) + A(u_r)\right)(u_\ell - u_r) + O(\varepsilon^2). \tag{5.30}$$

Similarly, we write approximations for $F^{(\alpha)}\left(\frac{1}{2}(u_\ell + u_r)\right), \alpha = 1, \cdots, d$, and deduce expressions for $[F^{(\alpha)}]$.

Substituting the above approximate relations in (4.109), and retaining only the leading order terms in $|u_\ell - u_r| = \varepsilon$, we get

$$\left\{\mathcal{S}_t \frac{1}{2}\left(A(u_\ell) + A(u_r)\right) + \mathcal{S}_{x_\alpha}\left(\frac{1}{2}(B^{(\alpha)}(u)_\ell + B^{(\alpha)}(u_r))\right)\right\}(u_\ell - u_r) = \mathbf{0}. \tag{5.31}$$

For a shock, $|u_\ell - u_r| \neq 0$. Let the approximate shock velocity corresponding to the approximate values of $[H]$ and $[F^{(\alpha)}]$ appearing in (5.31) be $S_{\ell r}$. Then the matrix

$$M_{\ell r} := -\frac{1}{2}\left(A(u_\ell) + A(u_r)\right) S_{\ell r} + \frac{1}{2}\left(B^{(\alpha)}(u_\ell) + \frac{1}{2}B^{(\alpha)}(u_r)\right) N_\alpha \tag{5.32}$$

is singular. Let $L_{\ell r}$ and $R_{\ell r}$ be the left and right eigenvectors of the matrix $M_{\ell r}$.

From (5.31) we get an approximate value of the jump in u

$$u_\ell - u_r = R_{\ell r} W, \quad W \neq 0, \tag{5.33}$$

where $W = O(\varepsilon)$ is a measure of the strength of the shock. Substituting (5.33) in (5.31), premultiplying by $L_{\ell r}$ and dividing by $\frac{1}{2}L_{\ell r}(A(u_\ell) + A(u_r))R_{\ell r}W$, we get a scalar result

$$\mathcal{S}_t + \left\{\frac{1}{2}\left(\frac{L_{\ell r}B^{(\alpha)}(u_\ell)R_{\ell r}}{\frac{1}{2}L_{\ell r}(A(u_\ell) + A(u_r))R_{\ell r}}\right) + \frac{1}{2}\left(\frac{L_{\ell r}B^{(\alpha)}(u_r)R_{\ell r}}{\frac{1}{2}L_{\ell r}(A(u_\ell) + A(u_r))R_{\ell r}}\right)\right\}\mathcal{S}_{x_\alpha} = 0. \tag{5.34}$$

Let c_ℓ and c_r be the velocities of the nonlinear wavefronts just behind and just ahead of the shock surface $S_t : \mathcal{S}(x, t) = 0$, respectively, and instantaneously coincident with it at time t. Let ℓ_ℓ and r_ℓ be the left and right eigenvectors of the matrix $M_\ell := -A(u_\ell)c_\ell + B^{(\alpha)}(u_\ell)N_\alpha$. Similarly, we define ℓ_r and r_r. Then, it is easy to show that $|c_l - S| = O(\varepsilon) = |c_r - S|$. We note

$$L_{\ell r} = \ell_\ell + O(\varepsilon),\ R_{\ell r} = r_\ell + O(\varepsilon),\ L_{\ell r} = \ell_r + O(\varepsilon),\ R_{\ell r} = r_r + O(\varepsilon). \tag{5.35}$$

Retaining only the most dominant terms, we can derive he following approximate results

$$\frac{\boldsymbol{L}_{\ell r}B^{(\alpha)}(\boldsymbol{u}_\ell)\boldsymbol{R}_{\ell r}}{\frac{1}{2}\boldsymbol{L}_{\ell r}(A(\boldsymbol{u}_\ell)+A(\boldsymbol{u}_r))\boldsymbol{R}_{\ell r}}=\frac{\boldsymbol{\ell}_\ell B^{(\alpha)}(\boldsymbol{u}_\ell)\boldsymbol{r}_\ell}{\boldsymbol{\ell}_\ell A(\boldsymbol{u}_\ell)\boldsymbol{r}_\ell}\ and\ \frac{\boldsymbol{L}_{\ell r}B^{(\alpha)}(\boldsymbol{u}_r)\boldsymbol{R}_{\ell r}}{\frac{1}{2}\boldsymbol{L}_{\ell r}(A(\boldsymbol{u}_\ell)+A(\boldsymbol{u}_r))\boldsymbol{R}_{\ell r}}=\frac{\boldsymbol{\ell}_r B^{(\alpha)}(\boldsymbol{u}_r)\boldsymbol{r}_r}{\boldsymbol{\ell}_r A(\boldsymbol{u}_r)\boldsymbol{r}_r}. \tag{5.36}$$

Finally, the approximate relation (5.34) is replaced by another approximate relation of (4.110) correct up to $O(\varepsilon)$ for a weak shock

$$Q_{\ell r}(t,\boldsymbol{x},\mathcal{S}_t,\mathcal{S}_{x_\alpha}):=\mathcal{S}_t+\frac{1}{2}\left\{\frac{\boldsymbol{\ell}_\ell B^{(\alpha)}(\boldsymbol{u}_\ell)\boldsymbol{r}_\ell}{\boldsymbol{\ell}_\ell A(\boldsymbol{u}_\ell)\boldsymbol{r}_\ell}+\frac{\boldsymbol{\ell}_r B^{(\alpha)}(\boldsymbol{u}_r)\boldsymbol{r}_r}{\boldsymbol{\ell}_r A(\boldsymbol{u}_r)\boldsymbol{r}_r}\right\}\mathcal{S}_{x_\alpha}=0 \tag{5.37}$$

which we shall like to treat as SME.

Since the functions $\boldsymbol{u}_\ell$ and $\boldsymbol{u}_r$ are defined only in the domains to the left and to the right of the shock, the relation (5.37) is valid only on the shock manifold $\mathcal{S}=0$ and not in a neighbourhood of it. Hence, it cannot be treated as a PDE. Assuming that $\boldsymbol{u}_\ell$ and $\boldsymbol{u}_r$ are smooth in their respective domains, we extend them on the other side of the shock as smooth functions. The relation (5.37) now becomes a PDE. The extensions are not unique, but since we extend them as smooth functions, the values of these functions and their derivatives are unique on $\mathcal{S}=0$ and this is sufficient for our purpose.

Note that $Q_\ell:=\mathcal{S}_t+\frac{\boldsymbol{\ell}_\ell B^{(\alpha)}(\boldsymbol{u}_\ell)\boldsymbol{r}_\ell}{\boldsymbol{\ell}_\ell A(\boldsymbol{u}_\ell)\boldsymbol{r}_\ell}\mathcal{S}_{x_\alpha}=0$ or $\mathcal{S}_t+(\chi_\alpha)_\ell\mathcal{S}_{x_\alpha}=0$ is the characteristic PDE (4.61) of the system (1.4) with $\boldsymbol{u}$ replaced by $\boldsymbol{u}_\ell$ and the normal $\boldsymbol{n}$ of the characteristic surface replaced by the normal $\boldsymbol{N}$ of the shock front. We can give a similar interpretation for the equation $Q_r:=\mathcal{S}_t+\frac{\boldsymbol{\ell}_r B^{(\alpha)}(\boldsymbol{u}_r)\boldsymbol{r}_\ell}{\boldsymbol{\ell}_r A(\boldsymbol{u}_r)\boldsymbol{r}_r}\mathcal{S}_{x_\alpha}=0$ or $\mathcal{S}_t+(\chi_\alpha)_r\mathcal{S}_{x_\alpha}=0$.

Since the velocity with components $\frac{\boldsymbol{\ell}B^{(\alpha)}\boldsymbol{r}}{\boldsymbol{\ell}A\boldsymbol{r}}$ of a nonlinear wavefront does satisfy the consistency condition (4.5), the sum $\frac{\boldsymbol{\ell}_\ell B^{(\alpha)}(\boldsymbol{u}_\ell)\boldsymbol{r}_\ell}{\boldsymbol{\ell}_\ell A(\boldsymbol{u}_\ell)\boldsymbol{r}_\ell}+\frac{\boldsymbol{\ell}_r B^{(\alpha)}(\boldsymbol{u}_r)\boldsymbol{r}_r}{\boldsymbol{\ell}_r (A(\boldsymbol{u}_r)\boldsymbol{r}_r}$ give components of a velocity which also satisfy it with $\boldsymbol{n}$ replaced by $\boldsymbol{N}$ and does qualify to be a ray velocity. Now we deduce from (5.37)

$$\frac{dX_\alpha}{dt}:=\left(\frac{dx_\alpha}{dt}\right)_{shock}=\frac{1}{2}\{(\chi_\alpha)_\ell+(\chi_\alpha)_r\}, \tag{5.38}$$

where $X(t)$ is the position of a point on the shock manifold. Since $\boldsymbol{u}_\ell$ and $\boldsymbol{N}$ appear in $(\chi_\alpha)_\ell$ in place of $\boldsymbol{u}$ and $\boldsymbol{n}$, $(\boldsymbol{\chi})_\ell$ is the ray velocity of the system (1.4) for a nonlinear wavefront just on the left the shock, provided we take the wavefront to be instantaneously coincident with the shock surface. Similar interpretation follows for $(\chi_\alpha)_r$. This equation proves the first part of the Theorem 5.3.1, namely the shock ray velocity $\frac{dX}{dt}$ is mean of the ray velocities of the nonlinear wavefronts just ahead and just behind the shock and instantaneously consistent with it.

We now give a derivation of the second part of the Theorem 5.3.1, namely

$$\frac{dN_\alpha}{dt} = \frac{1}{2}\{(\psi_\alpha)_\ell + (\psi_\alpha)_r\} \tag{5.39}$$

where $(\psi_\alpha)_\ell$ and $(\psi_\alpha)_r$ can be defined as $(\chi_\alpha)_\ell$ and $(\chi_\alpha)_r$ above. More explicitly $(\psi_\alpha)_\ell$ is the expression ψ_α in (4.63) with $\boldsymbol{u}$ and $\boldsymbol{n}$ replaced by $\boldsymbol{u}_\ell$ and $\boldsymbol{N}$, respectively, and represents the rate of turning of the nonlinear wavefront just behind the shock front and instantaneously coincident with it. A similar interpretation follows for $(\psi_\alpha)_r$. The proof is very simple and follows from the second part of the ray equations of SME (5.37) as (4.64) follows from (4.61).

The expressions for $(\psi_\alpha)_\ell$ and $(\psi_\alpha)_r$, contain operators $\left(\frac{\partial}{\partial\eta_\beta^\alpha}\right)_{shock}$. We note that the operators $\frac{d}{dt}$ in (5.38) and (5.39) and $\left(\frac{\partial}{\partial\eta_\beta^\alpha}\right)_{shock}$ appearing in $(\psi_\alpha)_\ell$ and $(\psi_\alpha)_r$ have explicit expressions:

$$\left(\frac{d}{dt}\right)_{shock} = \frac{\partial}{\partial t} + \left(\frac{1}{2}\{(\chi_\alpha)_\ell + (\chi_\alpha)_r\}\right)\frac{\partial}{\partial x_\alpha}, \quad \left(\frac{\partial}{\partial\eta_\beta^\alpha}\right)_{shock} = N_\beta\frac{\partial}{\partial x_\alpha} - N_\alpha\frac{\partial}{\partial x_\beta}. \tag{5.40}$$

These two Eqs. (5.38) and (5.39) complete the proof the theorem. ■

5.3.2 SRT in a Polytropic Gas from WNLRT

We take a forward facing shock front propagating into a polytropic gas, which is at rest and has constant values of pressure and density. This shock belongs to the characteristic field of c_5 in the set (4.53) and is followed by a one parameter family of forward facing nonlinear wavefronts belonging to the same characteristic field. These wavefronts move faster than the shock and hence catch up the shock, interact with it and then disappear. The short wave assumption for the perturbation containing wavefronts and the shock implies that the nonlinear wavefronts, interacting with the shock, will be instantaneously coincident with the shock. The ray equations of the WNLRT in three-space-dimensions for a particular nonlinear wavefront are (5.11). The unit normal to the shock front is $\boldsymbol{N}$. The instantaneously coincident nonlinear wavefront just ahead of the shock (this is actually a linear wavefront with $\tilde{w} = 0$) has the ray velocity: $\boldsymbol{n}$ multiplied by the local sound velocity a_0 and and hence has bicharacteristic velocity is $\boldsymbol{n}a_0$. We denote the amplitude $\tilde{w}$ the nonlinear wavefront just behind the shock and instantaneously coincident with it by $a_0\mu$. Then for the weak shock under consideration the non-dimensional shock amplitude is $\epsilon\mu$. For the instantaneously coincident nonlinear wave on the two sides of the shock, we have $\boldsymbol{n} = \boldsymbol{N}$. Using the theorem 5.3.1 and the results (5.11), we find that a point $\mathbf{X}$ on the shock ray satisfies

$$\frac{d\mathbf{X}}{dT} = \frac{1}{2}\left\{a_0\boldsymbol{N} + \boldsymbol{N}a_0\left(1+\epsilon\frac{\gamma+1}{2}\mu\right)\right\} = \mathbf{N}a_0\left(1+\epsilon\frac{\gamma+1}{4}\mu\right) \tag{5.41}$$

$$\frac{d\boldsymbol{N}}{dT} = -\frac{1}{2}\left\{0+\epsilon\frac{\gamma+1}{2}a_0\boldsymbol{L}_s\mu\right\} = -\epsilon\frac{\gamma+1}{4}a_0\boldsymbol{L}_s\mu \tag{5.42}$$

where T is the time measured while moving along a shock ray and $\boldsymbol{L}_s$ is the tangential operator $\nabla - \boldsymbol{N}\langle\boldsymbol{N},\nabla\rangle$ on the shock front $\mathcal{S}_t = 0$. With $\tilde{w} = a_0\mu$ and $\boldsymbol{n} = \boldsymbol{N}$ (5.12) becomes

$$\begin{aligned}\frac{d\mu}{dT} &\equiv \left\{\frac{\partial}{\partial t} + a_0\left(1+\epsilon\frac{\gamma+1}{4}\mu\right)\langle\boldsymbol{N},\nabla\rangle\right\}\mu \\ &= -\frac{1}{2}a_0\langle\nabla,\boldsymbol{N}\rangle\mu - \epsilon\frac{\gamma+1}{4}\mu\langle\boldsymbol{N},\nabla\rangle\tilde{w}\end{aligned} \tag{5.43}$$

Since μ is defined only on the shock front (and also on the instantaneously coincident nonlinear wavefront behind it but not on the other members of the one parameter family of wavefronts following it), the normal derivative of μ, i.e. $\langle\boldsymbol{N},\nabla\rangle\mu$ is not defined. We introduce new variables,[10] defined on the shock front:

$$\mu_1 = \epsilon\{\langle\boldsymbol{N},\nabla\rangle\tilde{w}\}\,|_{shockfront}, \quad \mu_2 = \epsilon^2\left\{\langle\boldsymbol{N},\nabla\rangle^2\tilde{w}\right\}|_{shockfront} \tag{5.44}$$

where powers of ϵ appears to make both μ_1 and μ_2 of $O(1)$ since, in short wave approximation, variation of $\tilde{w}$ with respect to the fast variable θ (introduced in (5.8)) is of the order of 1 and hence $< N, \nabla > \tilde{w} = O(\frac{1}{\epsilon})$ and $< N, \nabla >^2 \tilde{w} = O(\frac{1}{\epsilon^2})$.

The Eq. (5.43) leads to the first compatibility condition along a shock ray (first derived in [38])

$$\frac{d\mu}{dT} = a_0\Omega_s\mu - \frac{\gamma+1}{4}\mu\mu_1 \tag{5.45}$$

where

$$\Omega_s = -\frac{1}{2}\langle\nabla,N\rangle, \tag{5.46}$$

and the subscript s on Ω stands for the mean curvature of the shock. This is the energy transport equation along a shock ray.

To find the second compatibility condition along with a shock, we differentiate (5.12) in the direction of $\boldsymbol{n}$ but on the length scale over which θ varies. On this length scale, $\boldsymbol{n}$ may be approximately treated as constant and we get, after rearranging some terms,

---

[10] See (5.50) below, μ_1 has dimension of $\frac{a_0}{L}$ and μ_2 has dimension of $\frac{a_0^2}{L^2}$.

$$\left\{\frac{\partial}{\partial t}+\left(a_0+\epsilon\frac{\gamma+1}{4}\tilde{w}\right)\langle \boldsymbol{n},\nabla\rangle\right\}\langle \boldsymbol{n},\nabla\rangle\,\tilde{w}=-\frac{1}{2}a_0\langle\nabla,\boldsymbol{n}\rangle\langle\boldsymbol{n},\nabla\rangle\tilde{w}$$
$$-\epsilon\,\frac{\gamma+1}{2}\,\{\langle\boldsymbol{n},\nabla\rangle\tilde{w}\}^2-\epsilon\frac{\gamma+1}{4}\,\tilde{w}\,\langle\mathbf{n},\nabla\rangle^2\,\tilde{w} \quad (5.47)$$

Writing this equation on the wavefront instantaneously coincident with the shock (i.e setting $\boldsymbol{n}=\boldsymbol{N}$), multiplying it by ϵ we get

$$\frac{d\mu_1}{dT}=a_0\Omega_s\mu_1-\frac{\gamma+1}{2}\mu_1^2-\frac{\gamma+1}{4}a_0\mu\mu_2 \quad (5.48)$$

which is the second compatibility condition along shock rays given by (5.41) and (5.42). Similarly, higher order compatibility conditions can be derived.

Thus, for the Euler's equations, we have derived the infinite system of compatibility conditions for a weak shock just from the dominant terms of WNLRT. As we have already mentioned, the shock ray theory is an exact theory (weak shock assumption is another independent assumption) but since there are an infinite number of compatibility conditions on it, it is difficult, to use it for computing shock propagation. We now use the new theory of shock dynamics (NTSD) [49] according to which the system of Eqs. (5.41), (5.42), (5.45) and (5.48) can be closed by dropping the term containing μ_2 in the Eq. (5.48). This step is justified in the case $\mu_1 > 0$, which occurs very frequently in a applications such as a blast wave. But for multi-dimensional problems, dropping $\mu\mu_2$ gives excellent results in all cases [8]. When we consider the propagation of even stronger shocks in gas dynamics, the results of [42] in Chap. 8, show that neglecting the term $\mu\mu_2$ in the second compatibility condition gives good results for one space dimension for both accelerating and decelerating piston problems after an initial sudden push of the piston with a constant velocity.

Thus according to the new theory of shock dynamics, we drop the last term in (5.48) and get an equation for the gradient μ_1 in the form (first derived in [38])

$$\frac{d\mu_1}{dT}=a_0\Omega_s\mu_1-\frac{\gamma+1}{2}\mu_1^2. \quad (5.49)$$

5.4 Non-dimensional Form of Equations of WNRT and SRT

We non-dimensionalize the spatial coordinates $\boldsymbol{x}$ by a typical length L in the problem (say the radius of curvature of Ω_0 at a particular point), velocity by the sound velocity a_0 in the undisturbed state, time by L/a_0, curvatures Ω and Ω_s by $\frac{1}{L}$. We denote the non-dimensional spatial, temporal coordinates and curvatures again by the same symbols $\boldsymbol{x}$, t, Ω and Ω_s. We first note

$$\tilde{w} = \frac{a_0(\rho - \rho_0)}{\varepsilon \rho_0}, \quad \mu_1 = \epsilon\{\langle \boldsymbol{N}, \nabla\rangle \tilde{w}\}\mid_{shockfront}.$$

Now define a non-dimensional velocity m of the nonlinear wavefront, velocity M of shock front and the gradient of the flow behind of shock front $\mathcal{V}$ by

$$m = 1 + \varepsilon\frac{\gamma+1}{2a_0}\tilde{w} = 1 + \varepsilon\frac{\gamma+1}{2}\mu, \ M = 1 + \varepsilon\frac{\gamma+1}{4a_0}\tilde{w} = 1 + \varepsilon\frac{\gamma+1}{4}\mu, \ \mathcal{V} = \frac{(\gamma+1)L}{4a_0}\mu_1. \tag{5.50}$$

Equations of WNLRT: First, we note that the eikonal equation of the forward facing wave in WNLRT approximation is

$$\varphi_t + m|\nabla\varphi| = 0. \tag{5.51}$$

The equations (5.11)–(5.12) in non-dimensional variables become

$$\frac{d\boldsymbol{x}}{dt} = m\boldsymbol{n}, \quad \frac{d\boldsymbol{n}}{dt} = -\boldsymbol{L}m \tag{5.52}$$

and

$$\frac{dm}{dt} \equiv \{\frac{\partial}{\partial t} + m\langle \boldsymbol{n}, \nabla\rangle\}m = \Omega(m-1) \tag{5.53}$$

where $\boldsymbol{L}$ retains the same expression as in (4.72) or (5.13). An expanded form of this equation is

$$m_t + m\langle \boldsymbol{n}, \nabla\rangle m = -\frac{1}{2}\langle \nabla, \boldsymbol{n}\rangle(m-1). \tag{5.54}$$

It may be interesting to combine this equation with the eikonal equation (5.51) and develop a numerical method to find the evolution of a weakly nonlinear wavefront, may be following level set method.

Equations of SRT: The SRT Eqs. (5.41), (5.42), (5.45) and (5.49) in non-dimensional variables become

$$\frac{d\boldsymbol{X}}{dT} = M\boldsymbol{N}, \quad \frac{d\boldsymbol{N}}{dT} = -\boldsymbol{L}M, \tag{5.55}$$

$$\frac{dM}{dT} \equiv \{\frac{\partial}{\partial t} + M\langle \boldsymbol{N}, \nabla\rangle\}M = \Omega_s(M-1) - (M-1)\mathcal{V} \tag{5.56}$$

and

$$\frac{d\mathcal{V}}{dT} = \Omega_s\mathcal{V} - 2\mathcal{V}^2. \tag{5.57}$$

In the rest of the book by *SRT equations*, we shall not refer to an infinite system of equations, but the truncated system (5.55)–(5.57) as proposed in the NTSD.

We further mention that in the rest of the book, we shall use only the above non-dimensional equations of WNLRT and SRT.

Chapter 6
Kinematical Conservation Laws (KCL)

The general results[1] in this chapter may be a bit heavy and hence, a beginner to KCL may read the Sect. 8.1 in Chap. 8.

The weakly nonlinear ray theory and shock ray theory presented in Chap. 5, are in differential form and hence are valid as long as the fronts Ω_t remain smooth. When a discontinuity in the normal direction $\boldsymbol{n}$ and the wave amplitude w on Ω_t appears, the governing PDEs of Ω_t breakdown. Examples of Ω_t with discontinuities in $\boldsymbol{n}$ and w across a $(d-2)$ dimensional surface on Ω_t are plenty. The first theoretical evidence is in Whitham's 1957 and 1959 work (see [56]) on shock propagation, where he called the point of discontinuity on a shock front in 2-D and the curve of discontinuity on a shock front in 3-D '**shock–shock**'. Experimental evidence of such a discontinuity on a shock front in gas dynamics was first shown by Sturtevant and Kulkarni [54]. In a numerical computation of successive positions of a shock front by the shock ray theory of the Sect. 5.4, Kevlahan studied in 1996 the formation and propagation of shock–shocks by a special method [26], which is difficult to continue for a long time. In this chapter, we develop and analyze a mathematical theory to study formation and propagation of such discontinuities on an arbitrary surface Ω_t, which evolves in its own dynamics, and show that these discontinuities are images on physical space of shocks of a system of conservation laws (namely **kinematical conservation laws (KCL)**)in a ray coordinate system. Since these discontinuities can appear not only on a shock front but also on any moving surface Ω_t, we call them **kinks**.

Ray coordinate system: Consider the evolution of a $(d-1)$-dimensional surface Ω_t in $\mathbb{R}^d$. On Ω_t let $\boldsymbol{\xi} = (\xi_1, \xi_2, \ldots, \xi_{d-1})$ be a set of surface coordinates, which also evolve with time t (see Fig. 6.1 for $d = 3$). The surface Ω_t in $\boldsymbol{x}$-space is generated by a $d-2$ parameter family of curves such that along each of these curves a particularcoordinate ξ_p varies and the parameters $\xi_1, \ldots, \xi_{p-1}, \xi_{p+1}, \ldots, \xi_{d-1}$ are

[1]These results are reproduced from [46] with some changes. The author thanks Indian National Science Academy for kindly giving permission for reproduction.

P. Prasad, *Propagation of Multidimensional Nonlinear Waves and Kinematical Conservation Laws*, Infosys Science Foundation Series,
https://doi.org/10.1007/978-981-10-7581-0_6

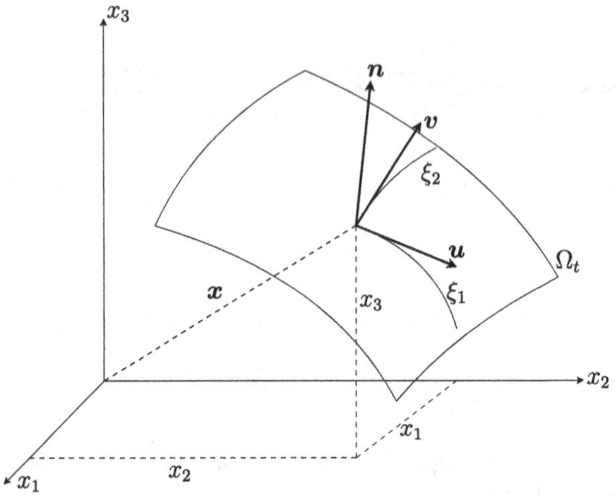

Fig. 6.1 A ray coordinate system of a surface Ω_t. This figure have been reproduced from [2] - Fig. 2.1 with permission from the Managing Editor, SIAM

constant.[2] This is true for each $p = 1, 2, \ldots, d-1$. Through each point $\boldsymbol{\xi}$ on Ω_t there passes a ray (in $\boldsymbol{x}$-space) associated with the evolution of Ω_t. Let the ray velocity be $\boldsymbol{\chi}$. The rays from different points of Ω_t are governed by the Eqs. (4.66)–(4.67) and form a $d-1$ parameter family of curves. Given a time t, we know the position Ω_t of the moving surface and given $\boldsymbol{\xi}$ we get a unique point $\boldsymbol{x}$ on Ω_t. Similarly, given a point $\boldsymbol{x} \in \mathbb{R}^d$, we know the time t when Ω_t arrives here and there is a value of $\boldsymbol{\xi}$ at this point. We assume that the mapping between the **ray coordinates** $(\boldsymbol{\xi}, t)$ and the spatial coordinates $\boldsymbol{x} = (x_1, x_2, \ldots, x_d)$ is locally one-to-one. This is true as long as the $d-1$ tangent vectors of the coordinate curves on Ω_t are linearly independent. Thus, given a domain in the physical space, we have two sets of coordinate systems, physical coordinates $\boldsymbol{x}$ and ray coordinates $(\boldsymbol{\xi}, t)$.

Metric associated with a coordinate ξ_p and that with t: Let the metric associated with the surface coordinate ξ_p be g_p. Then $g_p d\xi_p$ (no sum over the repeated subscript here) is an element of distance along the coordinate line along which ξ_p varies. The speed of a point moving along the ray with the ray velocity $\boldsymbol{\chi}$ is $|\boldsymbol{\chi}|$. Then, while moving with the ray velocity, $|\boldsymbol{\chi}|dt$ is the displacement along the ray in time dt. Thus, $|\boldsymbol{\chi}|$ is the metric associated with the coordinate t. When $\boldsymbol{\chi}$ is in the direction of the normal to Ω_t, as assumed below, the metric associated with t is velocity m of the surface.[3]

[2]As mentioned at end of the Chap. 1, the range of a repeated subscripts or superscript p and q for summation in a term in a will be $1, 2, \ldots, d-1$.

[3]Velocity of the surface is the velocity in the normal direction.

Historical notes: The 2-D kinematical conservation laws (KCL) were found by Morton, Prasad and Ravindran [33] first by manipulating the 2-D WNLRT equations[4] and soon after that by a systematic geometrical approach presented here. The paper was accepted by an associate editor of the European Journal of Applied Mathematics but then the Editor-in-Chief kept on insisting on repeated revisions and the authors decided not to pursue its publication. Subsequent papers based on it were published in many journals, mainly Journal of Fluid Mechanics. The book [42] contains an extension of of 2-D KCL to an anisotropic motion of a curve and a complete discussion. The 3-KCL were found by Giles, Prasad and Ravindran [20]. The manuscript of this work received good reviews but there were also questions, which could not be answered at that time. The theory of 3-D KCL was completed by Arun and Prasad, the results were published in two papers [3] (2009) and [4] (2010). The material in the chapter is an extension of the work in [3] to KCL in a space of arbitrary dimensions.

6.1 d-Dimensional KCL

As in the case of the 2-D KCL (8.6)–(8.7), discussed later, it is possible to formulate the d-D KCL system, when $\boldsymbol{\chi}$ is not orthogonal to Ω_t. But for simplicity, we consider here, the case when the motion of the surface Ω_t is **isotropic** in the sense that the associated ray velocity $\boldsymbol{\chi}$ depends on the unit normal $\boldsymbol{n}$ by the relation

$$\boldsymbol{\chi} = m\boldsymbol{n}, \tag{6.1}$$

where m, independent of $\boldsymbol{n}$, is the velocity of the surface. An example of this is the ray velocity of the wave equation $u_{tt} - m^2\left(u_{x_1x_1} + u_{x_2x_2} + \cdots + u_{x_dx_d}\right) = 0$, where m in general depends on $\boldsymbol{x}$ and t. For (6.1), the ray equations (4.6) and (4.7) take simple forms

$$\frac{d\boldsymbol{x}}{dt} = m\boldsymbol{n}, \quad |\boldsymbol{n}| = 1, \tag{6.2}$$

$$\frac{d\boldsymbol{n}}{dt} = -\boldsymbol{L}m := -\left(\nabla - \boldsymbol{n}\langle \boldsymbol{n}, \nabla\rangle\right)m, \tag{6.3}$$

where, the operator $\boldsymbol{L}$ is defined in (4.72). Since $|\boldsymbol{n}| = 1$, only $d-1$ equations in (6.3) are independent.

Let $\boldsymbol{u}_1, \boldsymbol{u}_2, \ldots, \boldsymbol{u}_{d-1}$ be unit vectors along the coordinates $\xi_1, \xi_2, \ldots, \xi_{d-1}$. Note that, like $\boldsymbol{n}$, $\boldsymbol{u}_p \in \mathbb{R}^d$. The unit vector along the ray, i.e. $\boldsymbol{n}$ is in the direction normal to the surface Ω_t and hence $\langle \boldsymbol{u}_p, \boldsymbol{n}\rangle = 0$.

[4]By chance and hence, there was no guaranty that they were physically realistic. We first derived the Eq. (8.16) and then manipulated them to derive (8.17).

Increments $d\xi_1, d\xi_2, \ldots, d\xi_{d-1}$ and dt in the ray coordinates leads to a displacement $d\boldsymbol{x}$ in $\boldsymbol{x}$-space given by

$$d\boldsymbol{x} = (g_1\boldsymbol{u}_1)d\xi_1 + (g_2\boldsymbol{u}_2)d\xi_2 + \cdots + (g_{d-1}\boldsymbol{u}_{d-1})d\xi_{d-1} + (m\boldsymbol{n})dt. \tag{6.4}$$

When the moving surface Ω_t is smooth, we equate $\boldsymbol{x}_{\xi_p t} = \boldsymbol{x}_{t\xi_p}$ and get the d-D KCL system

$$(g_p\boldsymbol{u}_p)_t - (m\boldsymbol{n})_{\xi_p} = 0, \ \textit{no sum over } p. \tag{6.5}$$

We also equate $\boldsymbol{x}_{\xi_p\xi_q}$ and $\boldsymbol{x}_{\xi_q\xi_p}$ to derive $(d-2)!$ more vector equations:

$$(g_q\boldsymbol{u}_q)_{\xi_p} - (g_p\boldsymbol{u}_p)_{\xi_q} = 0, \ \textit{no sum over repeated subscripts.} \tag{6.6}$$

Theorem 6.1.1 *For a smooth Ω_t Eq. (6.5) imply that $(g_q\boldsymbol{u}_q)_{\xi_p} - (g_p\boldsymbol{u}_p)_{\xi_q}$ is independent of t.*

Proof Differentiate (6.5) with respect to ξ_q and (6.5), with p replaced by q, with respect to ξ_p and subtract the second result from the first to get

$$\left((g_q\boldsymbol{u}_q)_{\xi_p} - (g_p\boldsymbol{u}_p)_{\xi_q}\right)_t = 0, \ \textit{no sum over repeated subscripts.}$$

This completes the proof of the theorem. □

When coordinates $\xi_1, \xi_2, \ldots, \xi_{d-1}$ are chosen on Ω_0, i.e. at $t = 0$, the expression for a displacement $d\boldsymbol{x}$ on Ω_0 will be given by (6.4) with $dt = 0$. Hence, the conditions (6.6) will be automatically satisfied at $t = 0$. The theorem implies that the Eq. (6.6) are satisfied at all $t > 0$. The Eq. (6.6) simply appear as constraints, just as a solenoidal condition in the equations of magnetohydrodynamics. Hence, we call the Eq. (6.6) as a **geometrical solenoidal constraint**.

KCL system: For $p = 1, 2, \ldots, d-1$, (6.5) gives a KCL system of $d-1$ vector equations, each one with d components. Thus, we have $d(d-1)$ scalar equations in (6.5). The KCL consists of just $d(d-1)$ conservation laws.

Remark 6.1.2 Once we have chosen the coordinates $\boldsymbol{\xi} = (\xi_1, \xi_2, \ldots, \xi_{d-1})$ on Ω_0, they remain coordinates on Ω_t, $\forall\, t > 0$ as if they are glued to it. Their directions $\boldsymbol{u}_1, \boldsymbol{u}_2, \ldots, \boldsymbol{u}_{d-1}$ evolve with time as the surface Ω_t evolves. We may initially choose an orthogonal coordinate system on Ω_0 but for $t > 0$ the system, in almost all cases, will not remain an orthogonal.[5]

KCL system is under-determined: Each the unit vectors $\boldsymbol{n}, \boldsymbol{u}_1, \boldsymbol{u}_2, \ldots \boldsymbol{u}_{p-1}$ has d components, out of which only $d-1$ are independent. These vectors satisfy $d-1$

[5]We give three simple examples of Ω_t, when orthogonality is not lost: (i) plane, (ii) a circular cylinder and (iii) a sphere, all three with suitable choices of surface coordinates and with constant distributions of m at $t = 0$. If the orthogonality is lost, it becomes very difficult to find the eigenvalues and discuss the nature of the eigenspace of a system of equations in KCL-based applications as one can see in references [3].

linear homogeneous equations $\langle \boldsymbol{u}_p, \boldsymbol{n} \rangle = 0$, $p = 1, 2, \ldots, d-1$; from which we can uniquely solve $\boldsymbol{n}$ in terms of the $d-1$ vectors $\boldsymbol{u}_p$. Therefore, the unknowns appearing in the system (6.5) are:

- $g_1, g_2, \ldots, g_{d-1}$, which are $d-1$,
- m, which is just one and
- $\boldsymbol{u}_1, \boldsymbol{u}_2, \ldots, \boldsymbol{u}_{d-1}$, have $(d-1)^2$ independent quantities since each $|\boldsymbol{u}_p| = 1$.

Thus, we have $d(d-1)+1$ unknowns in $d(d-1)$ Eq. (6.5) and hence, the KCL system is under-determined. This is expected, as we have taken a propagating front Ω_t without prescribing the physical nature of the front. The system can be closed only with the help of additional relations or equations, which would follow from the nature of the surface Ω_t, i.e. the dynamics of the medium in which it propagates.

Differential form of KCL: We shall now derive the differential form of the KCL system (6.5) assuming that $\boldsymbol{u}_p$, $\boldsymbol{n}$, m and g_p are smooth functions. We first note that $\langle \boldsymbol{u}_p, \boldsymbol{n} \rangle = 0$ implies $\langle \boldsymbol{u}_p, \boldsymbol{n}_{\xi_q} \rangle = -\langle (\boldsymbol{u}_p)_{\xi_q}, \boldsymbol{n} \rangle$. Further, since $|\boldsymbol{u}_p| = 1$, we have $\langle \boldsymbol{u}_p, (\boldsymbol{u}_p)_t \rangle = 0$. Writing the differential form of (6.5) and taking inner product with $\boldsymbol{u}_p$, we get

$$(g_p)_t = -m\langle \boldsymbol{n}, (\boldsymbol{u}_p)_{\xi_p} \rangle, \quad \textit{no sum over } p. \tag{6.7}$$

In the differential form of (6.5), we now use the expression for $(g_p)_t$ from (6.7) to get

$$g_p(\boldsymbol{u}_p)_t = m_{\xi_p}\boldsymbol{n} + m\langle \boldsymbol{n}, (\boldsymbol{u}_p)_{\xi_p} \rangle \boldsymbol{u}_p + m\boldsymbol{n}_{\xi_p}, \quad \textit{no sum over } p. \tag{6.8}$$

The total number of equations in (6.7) and (6.8) are $d + d(d-1) = d^2$, but we shall show that only $d(d-1)$ are independent. We prove a theorem:

Theorem 6.1.3 *For any given p, the last of the equations in (6.8), i.e. the dth equation*

$$g_p(u_{pd})_t = m_{\xi_p} n_d + m\langle \boldsymbol{n}, (\boldsymbol{u}_p)_{\xi_p} \rangle u_{pd} + m(n_d)_{\xi_p}, \quad \textit{no sum over } p \tag{6.9}$$

follow from its first d − 1 equations

$$g_p(u_{pq})_t = m_{\xi_p} n_q + m\langle \boldsymbol{n}, (\boldsymbol{u}_p)_{\xi_p} \rangle u_{pq} + m(n_q)_{\xi_p}, \textit{ no sum over} p;\ q = 1, 2, \ldots, d-1. \tag{6.10}$$

Proof We multiply (6.10) by u_{pq} and sum over q on the range $1, 2, \ldots, d-1$ to get

$$\begin{aligned}\frac{1}{2} g_p\{(u_{p1})^2 + (u_{p2})^2 + \cdots + (u_{p(d-1)})^2\}_t &= m_{\xi_p}\{n_1 u_{p1} + n_2 u_{p2} + \cdots + n_{(d-1)} u_{p(d-1)}\} \\ &+ m\langle \boldsymbol{n}, (\boldsymbol{u}_p)_{\xi_p} \rangle \{u_{p1}^2 + u_{p2}^2 + \cdots + u_{p(d-1)}^2\} + m\{u_{p1}(n_1)_{\xi_p} + u_{p2}(n_2)_{\xi_p} + \cdots \\ &+ u_{p(d-1)}(n_{d-1})_{\xi_p}\}, \quad \textit{no sum over } p.\end{aligned} \tag{6.11}$$

Now we use, without summation convention,

$$
\begin{aligned}
&(u_{p1})^2 + (u_{p2})^2 + \cdots + (u_{p(d-1)})^2 = 1 - (u_{pd})^2, \\
&n_1 u_{p1} + n_2 u_{p2} + \cdots + n_{(d-1)} u_{p(d-1)} = -n_d u_{pd}, \\
&u_{p1}(n_1)_{\xi_p} + u_{p2}(n_2)_{\xi_p} + \cdots + u_{p(d-1)}(n_{d-1})_{\xi_p} \\
&\qquad = \langle \boldsymbol{u}_p, \boldsymbol{n}_{\xi_p} \rangle - u_{pd}(n_d)_{\xi_p} = -\langle (\boldsymbol{u}_p)_{\xi_p}, \boldsymbol{n} \rangle - u_{pd}(n_d)_{\xi_p}
\end{aligned}
$$

and divide the resulting equation by u_{pd} to get (6.9).

We express u_{1d} in terms of $u_{11}, u_{12}, \ldots, u_{1(d-1)}$ using $u_{1d}^2 = 1 - (u_{11}^2 + u_{12}^2 + \cdots + u_{1(d-1)}^2)$ and do this for the last components of rest of the vectors $\boldsymbol{u}_2, \ldots, \boldsymbol{u}_{(d-1)}$. Then, we have expressed $u_{1d}, u_{2d}, \ldots, u_{(d-1)d}$ in terms of $u_{p1}, u_{p2}, \ldots, u_{p(d-1)}$; $p = 1, 2, \ldots, d-1$. Now the Eqs. (6.7) and (6.10) are a set of $d(d-1)$ independent differential form of the KCL system containing the $d(d-1)+1$ quantities m; g_p; $u_{p1}, u_{p2}, \ldots, u_{p(d-1)}$; $p = 1, 2, \ldots, d-1$.

6.2 Equivalence of KCL and Ray Equations

The equivalence of the KCL system for smooth solutions and the ray equations is not surprising as both are related geometrically. This equivalence is to be shown between ray equations (6.2) and (6.3); and the differential forms (6.7) and (6.8) of the KCL. **In this section, there is no sum over** p.

Theorem 6.2.1 *For a given smooth function m of $\boldsymbol{x}$ and t, the ray equations (6.2) and (6.3) are equivalent to the KCL as long as their solutions are smooth.*

Proof Let us first give an explicit derivation of the KCL system from the ray equations.

Derivation of (6.7): (6.4) gives $\boldsymbol{x}_{\xi_p} = g_p \boldsymbol{u}_p$ which implies $g_p^2 = x_{1\xi_p}^2 + x_{2\xi_p}^2 + \cdots + x_{d\xi_p}^2 = |\boldsymbol{x}_{\xi_p}|^2$. Differentiating it with respect to t, using $\boldsymbol{x}_{\xi_p t} = \boldsymbol{x}_{t\xi_p} \equiv (\boldsymbol{x}_t)_{\xi_p}$, $\boldsymbol{x}_{\xi_p} = g_p \boldsymbol{u}_p$ and dividing by g_p we get

$$(g_p)_t = \langle \boldsymbol{u}_p, (\boldsymbol{x}_t)_{\xi_p} \rangle. \tag{6.12}$$

Using (6.2) we derive

$$(g_p)_t = \langle \boldsymbol{u}_p, m_{\xi_p} \boldsymbol{n} + m \boldsymbol{n}_{\xi_p} \rangle = \langle \boldsymbol{u}_p, m \boldsymbol{n}_{\xi_p} \rangle = -m \langle \boldsymbol{n}, (\boldsymbol{u}_p)_{\xi_p} \rangle, \tag{6.13}$$

which is the Eq. (6.7).

Derivation of (6.8): We differentiate with respect to t the relation $g_p \boldsymbol{u}_p = \boldsymbol{x}_{\xi_p}$, and use $\boldsymbol{x}_{\xi_p t} = \boldsymbol{x}_{t\xi_p} = (m\boldsymbol{n})_{\xi_p}$ and use $g_{p_t} = -m\langle \boldsymbol{n}, (\boldsymbol{u}_p)_{\xi_p} \rangle$ to give

$$g_p(\boldsymbol{u}_p)_t = m\langle \boldsymbol{n}, (\boldsymbol{u}_p)_{\xi_p}\rangle \boldsymbol{u}_p + \boldsymbol{n} m_{\xi_p} + m\boldsymbol{n}_{\xi_p}, \tag{6.14}$$

which is the Eq. (6.8).
We give an alternative geometrical proof:
This proof is beautiful and equally simple. Let m be a smooth function of $\boldsymbol{x}$ and t and let $\boldsymbol{x}, \boldsymbol{n}$ (with $|\boldsymbol{n}| = 1$) satisfy the ray equations (6.2) and (6.3), which (as described in Sect. 4.1.1) give successive positions of a moving surface Ω_t. Choose a coordinate system $\xi_1, \xi_2, \ldots, \xi_{p-1}$ on Ω_t with metrics $g_1, g_2, \ldots, g_{p-1}$ associated with these coordinates. Let $(\boldsymbol{u}_1, \boldsymbol{u}_2, \ldots, \boldsymbol{u}_{p-1})$ be unit tangent vectors along the coordinates curves. Then, the derivation given in the beginning of the last section leads to the d-D KCL system (6.5) along with the geometrical solenoidal constraint (6.6). Thus the ray equations imply d-D KCL.

KCL implies ray equations: We take d smooth linearly independent unit vector fields $\boldsymbol{u}_1, \boldsymbol{u}_2, \ldots, \boldsymbol{u}_{d-1}, \boldsymbol{n}$ and d smooth scalar functions $g_1, g_2, \ldots, g_{d-1}, m$ in $(\xi_1, \xi_2, \ldots, \xi_{d-1}, t)$-space such that $\boldsymbol{n}$ is orthogonal to each of $\boldsymbol{u}_1, \boldsymbol{u}_2, \ldots, \boldsymbol{u}_{d-1}$; and they satisfy the KCL (6.5) and the geometrical solenoidal constraint (6.6).

According to the fundamental integrability theorem ([15]-page 104), the conditions (6.5) and (6.6) imply the existence of a vector $\boldsymbol{x}$ satisfying

$$(\boldsymbol{x}_t, \boldsymbol{x}_{\xi_1}, \ldots, \boldsymbol{x}_{\xi_{d-1}}) = (m\boldsymbol{n}, g_1\boldsymbol{u}_1, \ldots, g_{d-1}\boldsymbol{u}_{d-1}). \tag{6.15}$$

Since the vectors on the right-hand side are linearly independent, this gives a local one to one mapping between $\boldsymbol{x}$-space and $(\xi_1, \xi_2, \ldots, \xi_{p-1}, t)$-space. Let the plane $t =$ constant in $(\xi_1, \xi_2, \ldots, \xi_{p-1}, t)$-space be mapped on to a surface Ω_t in $\boldsymbol{x}$-space on which $\xi_1, \xi_2, \ldots, \xi_{p-1}$ are surface coordinates. Then $\boldsymbol{u}_1, \boldsymbol{u}_2, \ldots, \boldsymbol{u}_{p-1}$ are tangent vectors to Ω_t and $\boldsymbol{n}$ is orthogonal to Ω_t, i.e. it is a unit normal vector of Ω_t. Let $\varphi(\boldsymbol{x}, t) = 0$ be the equation of Ω_t, then $\boldsymbol{n} = \nabla\varphi/|\nabla\varphi|$. The relation $\boldsymbol{x}_t = m\boldsymbol{n}$ in (6.15) is nothing but the first part of the ray equation. The normal velocity of Ω_t is m which must be equal to $-\varphi_t/|\nabla\varphi|$. The function φ now satisfies the eikonal equation $\varphi_t + m|\nabla\varphi| = 0$, which implies (6.3), see also [43]. Thus, we have derived the ray equations from KCL. Now we have completed the proof of the theorem. □

6.3 Kink Phenomenon

In the derivation of KCL and discussion of some of its properties, we have taken the functions appearing in the equations to be smooth and the transformation between the ray coordinates $(\boldsymbol{\xi}, t)$ and physical coordinates $\boldsymbol{x}$ be nonsingular. When a singularity in the transformation appears, the shape of a surface Ω_t would become very complex in multi-dimensions, and probably a complete theory is not available. Some discussion for the 2-D case is available in Sect. 30 [14]. In this section, we do not discuss singularities due to the vanishing of the Jacobian of the transformation but

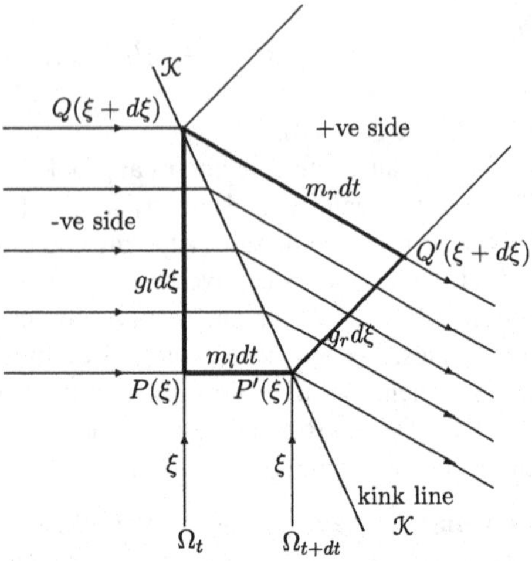

Fig. 6.2 A 2-D analogue of the kink phenomenon corresponding to a backward moving shock, i.e. moving in the negative ξ direction, here $d\xi > 0$. In this case, Ω_t is a curve consisting of two lines meeting at a kink $\mathcal{K}_t$, which we have shown by a point Q on Ω_t. The kink surface $\mathcal{K}$ becomes a kink line passing through the points Q and P'

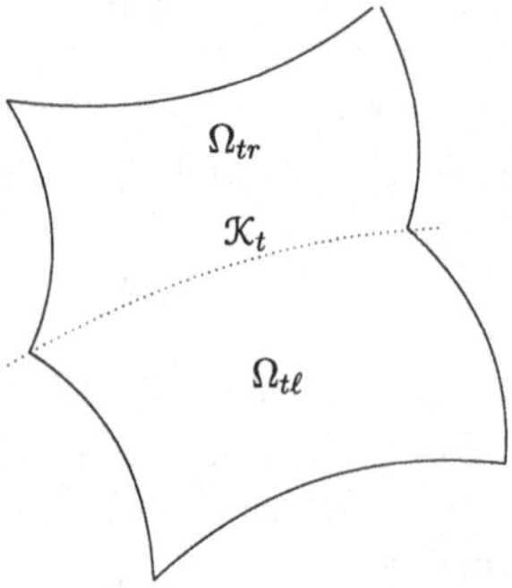

Fig. 6.3 Figure is drawn in three space dimensions. Kink $\mathcal{K}_t$ (shown with dotted lines) on $\Omega_t = \Omega_{tr} \cup \Omega_{t\ell}$. This figure have been reproduced from [2] - Fig. 2.2 with permission from the Managing Editor, SIAM

devote to images on $\boldsymbol{x}$-space of a special type of singularity in functions in the ray coordinates (Sect. 3.3.3 of [42]).

The system (6.5) consists of equations which are conservation laws, so its weak solution may contain a **shock surface** represented $\mathbb{S}$, which is a $d-1$ dimensional surface in $(\boldsymbol{\xi}, t)$-space.[6] Across this shock surface the d scalars m, g_p and $d-1$ vectors $\boldsymbol{u}_p$; and $\boldsymbol{n}$ will be discontinuous and will satisfy RH relations. Image of a shock surface $\mathbb{S}$ into $\boldsymbol{x}$-space will be another $d-1$ dimensional surface, let us call it a **kink surface** and denote it by $\mathcal{K}$. The surface $\mathcal{K}$ will intersect Ω_t in a $d-2$ dimensional surface, say **kink** which we denote by $\mathcal{K}_t$. In order to avoid using 'kink surface' for both $\mathcal{K}$ and $\mathcal{K}_t$, we call the later one simply 'kink'. In 2-D a kink is a point (Fig. 6.2), in 3-D it is a curve (Fig. 6.3) and in d-D it is a $d-2$ dimensional surface. As time t evolves, $\mathcal{K}_t$ will generate the kink surface $\mathcal{K}$. When a ray (which is a curve) crosses the kink at time t, the direction of the ray or normal direction $\boldsymbol{n}$

[6] A **shock front** (a phrase very commonly used in the literature) is a $d-2$ dimensional surface $\mathbb{S}_t$ in $d-1$ dimensional $\boldsymbol{\xi}$-space and its motion as t varies generates the shock surface $\mathbb{S}$ in $(\boldsymbol{\xi}, t)$-space.

of Ω_t jumps as seen in the Figs. 6.2 and 6.3. We assume that the mapping between $(\boldsymbol{\xi}, t)$-space and $\boldsymbol{x}$-space continues to be one to one even when a kink appears.

RH relations and conservation of distance: Consider any two points $P'(\boldsymbol{x})$ and $Q(\boldsymbol{x}+d\boldsymbol{x})$ in the physical space corresponding to points $(\boldsymbol{\xi}, t+dt)$ and $(\boldsymbol{\xi}+d\boldsymbol{\xi}, t)$ in the ray coordinate space. The distance between P' and Q is given by (6.4). Let us assume that Q and P' lie on the surfaces Ω_t and Ω_{t+dt} respectively[7] and we further assume that they lie on a kink surface $\mathcal{K}$, which intersects Ω_t and Ω_{t+dt} respectively in kinks $\mathcal{K}_t$ and $\mathcal{K}_{t+dt}$. Now $(\boldsymbol{\xi}, t+dt)$ and $(\boldsymbol{\xi}+d\boldsymbol{\xi}, t)$ lie on the shock surface $\mathbb{S}$ in $(\boldsymbol{\xi}, t)$-space. In Fig. 6.2, we present a 2-D analogue of the kink phenomenon.

We express the segment $QP' = d\boldsymbol{x}$ in terms of the quantities in the states on the negative and positive sides of the $\mathcal{K}$ denoted by subscripts ℓ and r, respectively. The conservation of $d\boldsymbol{x}$ implies that the expression for $(d\boldsymbol{x})_\ell$ on one side of the kink surface must be equal to that for $(d\boldsymbol{x})_r$, i.e.

$$\begin{aligned} d\boldsymbol{x} &= (g_{1\ell}\boldsymbol{u}_{1\ell})d\xi_1 + (g_{2\ell}\boldsymbol{u}_{2\ell})d\xi_2 + \cdots + (g_{(d-1)\ell}\boldsymbol{u}_{(d-1)\ell})d\xi_{d-1} + (m_\ell\boldsymbol{n}_\ell)dt \\ &= (g_{1r}\boldsymbol{u}_{1r})d\xi_1 + (g_{2r}\boldsymbol{u}_{2r})d\xi_2 + \cdots + (g_{(d-1)r}\boldsymbol{u}_{(d-1)r})d\xi_{d-1} + (m_r\boldsymbol{n}_r)dt. \end{aligned} \tag{6.16}$$

We have assumed $QP' = d\boldsymbol{x}$ on the kink surface $\mathcal{K}$ and its image $(d\boldsymbol{\xi}, dt)$ in ray coordinate space is on the shock surface $\mathbb{S}$. Projection of $(d\boldsymbol{\xi}, dt)$ on $\boldsymbol{\xi}$-space need not be in a direction normal to the shock front $\mathbb{S}_t$. We take the direction of the line element QP' in $\boldsymbol{x}$-space such that this projection on $\boldsymbol{\xi}$-space is in the direction of the normal to the shock front in $\boldsymbol{\xi}$-space, then the differentials are further restricted. Let the unit normal of the shock front $\mathbb{S}_t$ be $\boldsymbol{E} = (E_1, E_2, \ldots, E_{d-1})$ in $\boldsymbol{\xi}$-space and let the scalar S be the velocity of propagation of $\mathbb{S}_t$ in this space, then the differentials in (6.16) satisfy $\frac{d\xi_p}{dt} = E_p S$ and (6.16) now becomes

$$\begin{aligned} &(g_{1\ell}E_1\boldsymbol{u}_{1\ell} + g_{2\ell}E_2\boldsymbol{u}_{2\ell} + \cdots + g_{(d-1)\ell}E_{d-1}\boldsymbol{u}_{(d-1)\ell})S + m_\ell\boldsymbol{n}_\ell \\ &\quad = (g_{1r}E_1\boldsymbol{u}_{1r} + g_{2r}E_2\boldsymbol{u}_{2r} + \cdots + g_{(d-1)r}E_{d-1}\boldsymbol{u}_{(d-1)r})S + m_r\boldsymbol{n}_r. \end{aligned} \tag{6.17}$$

Remark 6.3.1 From the discussion above it follows that (6.17) is a condition for the conservation of distance across a kink $\mathcal{K}_t$ when we choose $\frac{d\boldsymbol{\xi}}{dt} = \boldsymbol{E}S$. We note that each of $\boldsymbol{u}_p$ is a vector with d components, therefore, the conservation of distance refers to conservation in all d coordinate directions and hence in any direction in physical $\boldsymbol{x}$-space.

The jump conditions or the RH relations across a shock surface $\mathbb{S}$ in a solution of the conservation laws (6.5) are

[7] For a 3-D analogue need to see the Fig. 6.3, where we have drawn only Ω_t, which has two parts Ω_{tr} and $\Omega_{t\ell}$ separated kink $\mathcal{K}_t$. Imagine Ω_{t+dt} above Ω_t. $\mathcal{K}_t$, when moves in time, will generate $\mathcal{K}$. Two points one on Ω_t and another on Ω_{t+dt} need not be on $\mathcal{K}$ and we make the second assumption. Further, since $\mathbb{S}$ is surface in 3-d (ξ_1, ξ_2, t)-space, projection on $\boldsymbol{\xi}$-plane of the line joining $(\boldsymbol{\xi}, t+dt)$ and $(\boldsymbol{\xi}+d\boldsymbol{\xi}, t)$ need not be in a normal direction to $\mathbb{S}_t$ and we make the third assumption.

$$S[g_p \boldsymbol{u}_p] + E_p[m\boldsymbol{n}] = 0, \ \textit{no sum over } p; \quad p = 1, 2, \ldots, d-1. \tag{6.18}$$

Multiplying (6.18) by E_p, summing over the subscript p on its range $1, 2, \ldots, d-1$ and using $|\boldsymbol{E}| = 1$, we get

$$S\left(E_1[g_1\boldsymbol{u}_1] + E_2[g_2\boldsymbol{u}_2] + \cdots + E_{p-1}[g_{p-1}\boldsymbol{u}_{p-1}]\right) + [m\boldsymbol{n}] = 0, \tag{6.19}$$

which is the same as (6.17). Thus, we have proved an extension of a theorem of GPR [20] for 3-D space:

Theorem 6.3.2 *The $d(d-1)$ jump relations (6.18) imply conservation of distance in $x_1, x_2, \ldots, x_d$ directions (and hence in any arbitrary direction in $\boldsymbol{x}$-space) in the sense that the expressions for a vector displacement $(d\boldsymbol{x})_{\mathcal{K}_t}$ of a point of the kink $\mathcal{K}_t$ in an infinitesimal time interval dt, when computed in terms of variables on the two sides of a kink surface, have the same value.*

This theorem assures that the d-D KCL are physically realistic.

Chapter 7
Conservation Forms of Energy Transport Equations

In this chapter, we shall first derive the conservation forms of the energy transport equation (5.12) of WNLRT, and Eqs. (5.45) an(5.49) of the SRT first in 3 space dimensions and then as a particular case in 2 space dimensions. Note that, we use non-dimensional variables as explained in Sect. 5.4. WNLRT and SRT are theories for propagation of nonlinear wavefronts and shock fronts in a **polytropic gas**.

7.1 Ray Tube Area

To begin with, we take an anisotropic motion of a surface Ω_t for which the associated rays, with ray velocity χ, are not necessarily orthogonal to Ω_t. Consider, a ray tube formed by the rays starting from the boundary points of an elementary area S_0 of Ω_0, see Fig. 7.1. Let S_t be the area of a section of Ω_t truncated by the ray tube. Let S_t' be the cross-sectional areas of the ray tube at t, i.e. the area of the section by a plane orthogonal to the tube. Let ψ^t be the angle between χ and normal $\boldsymbol{n}$ of Ω_t, then $S_t' = S_t \cos\psi^t$. We define the ray tube area $\mathcal{A}$ by [42, 56]

$$\mathcal{A} := \lim_{S_0' \to 0} \frac{S_t'}{S_0'} = \lim_{S_0 \to 0} \frac{S_t \cos\psi^t}{S_0 \cos\psi^0}, \tag{7.1}$$

where $S_0 \to 0$ in such a way that its longest diameter tends to 0. We note $\mathcal{A}_0 = 1$.

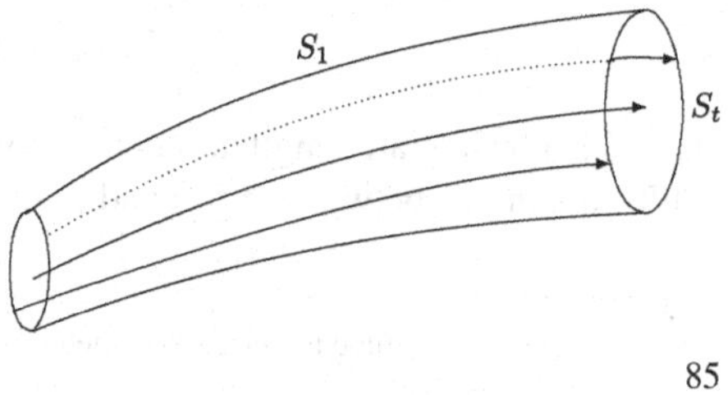

Fig. 7.1 Sketch of a ray tube reproduced from [2]

P. Prasad, *Propagation of Multidimensional Nonlinear Waves and Kinematical Conservation Laws*, Infosys Science Foundation Series,
https://doi.org/10.1007/978-981-10-7581-0_7

Theorem 7.1.1 *The ray tube area of the ray system is related to the divergence of the vector field* χ *by*

$$\frac{1}{\mathcal{A}}\frac{\partial \mathcal{A}}{\partial \ell} = \langle \nabla, \chi \rangle, \tag{7.2}$$

where ℓ *is the arc length along a ray, i.e.* $d\ell = |\chi| dt$.

Proof Let D be the domain bounded by sections S_0, S_t and the part S_1 between S_0 and S_t of a curved surface formed by the rays starting from the boundary points of S_0. Consider an integral $\int_D \langle \nabla, \chi \rangle dD$ over a volume D. Let us use the Gauss theorem

$$\int_D \langle \nabla, \chi \rangle dD = \int_{S_0} \langle \boldsymbol{n}_{S_0}, \chi \rangle dS + \int_{S_t} \langle \boldsymbol{n}_{S_t}, \chi \rangle dS + \int_{S_1} \langle \boldsymbol{n}_{S_1}, \chi \rangle dS. \tag{7.3}$$

Using $\langle \boldsymbol{n}_{S_0}, \chi \rangle = \cos \psi^0$ and $\langle \boldsymbol{n}_{S_t}, \chi \rangle = \cos \psi^t$ in the first two terms on the right-hand side and $\langle \boldsymbol{n}_{S_1}, \chi \rangle = 0$ in the last term we get

$$\int_D \langle \nabla, \chi \rangle dD = -\int_{S_0} \cos \psi^0 dS + \int_{S_t} \cos \psi^t dS = -S'_0 + S'_t. \tag{7.4}$$

For an elementary ray tube $\int_D \langle \nabla, \chi \rangle dD \approx \int_0^\ell \langle \nabla, \chi \rangle S' d\ell$. Now, we take the limit as $t \to 0$ (this also means $\ell \to 0$). This gives $\int_D \langle \nabla, \chi \rangle dD \approx S'_0 \langle \nabla, \chi \rangle_0 (\ell_t - \ell_0)$, where $\ell_0 = 0$. Finally, we get up to first-order terms

$$S'_0 \langle \nabla, \chi \rangle (\ell_t - \ell_0) = S'_t - S'_0. \tag{7.5}$$

In the limit,[1] when S_0 (or S_t) $\to 0$ and $d\ell = \ell_t - \ell_0 \to 0$, we get the exact result stated in the theorem. ■

7.2 Energy Transport Equation in 3-D WNLRT

For a weakly nonlinear wavefront Ω_t moving into an ambient gas at rest the rays satisfying (5.52) are orthogonal to Ω_t, so $\chi = \boldsymbol{n}m$, $\psi^t = 0$ and $S'_t = S_t$. Therefore, the ray tube area $\mathcal{A}$ of a weakly nonlinear wavefront is related to the mean curvature $\mathit{\Omega}$ (see (5.6)) by

$$\frac{1}{\mathcal{A}}\frac{\partial \mathcal{A}}{\partial \ell} = \langle \nabla, \boldsymbol{n} \rangle \equiv -2\mathit{\Omega}, \tag{7.6}$$

where ℓ is now the arc length along a ray of a nonlinear wavefront, i.e. $d\ell = m dt$.

Eliminating $\mathit{\Omega}$ between (5.53) and (7.6) we get

[1] For an elementary ray tube the largest diameter of a section S' at any ℓ tends to zero.

$$\frac{2mm_t}{m-1} = -\frac{1}{\mathcal{A}}\mathcal{A}_t, \tag{7.7}$$

which immediately leads to a conservation law for energy

$$\left\{(m-1)^2 e^{2(m-1)}\mathcal{A}\right\}_t = 0, \tag{7.8}$$

where the partial derivative with respect to t is in ray coordinates.

Let us compare the result (7.8) with the energy transport equation of the linear theory, where the non-dimensional value of a_0 is equal to 1. In linear theory, the eikonal equation and the transport equation (5.5) (here $u_1 = m - 1$) become

$$\varphi_t + |\nabla\varphi| = 1, \quad m_t = -\frac{1}{2\mathcal{A}}\mathcal{A}_t(m-1), \tag{7.9}$$

where in the second equation partial derivative with respect to t is the time derivative along a linear ray, i.e. $\frac{\partial}{\partial t} = \frac{\partial}{\partial t} + \langle \boldsymbol{n}, \nabla\rangle$. The Eq. (7.6) remains the same but now $d\ell = dt$ along the linear ray. The conservation form of the energy transport equation along the linear ray follows from the second equation in (7.9) as $\left\{(m-1)^2\mathcal{A}\right\}_t = 0$. Thus, the flux of energy density crossing a cross-section of a ray tube is proportional $(m-1)^2$, which is a well-known result [56], p. 245.

The ray tube area $\mathcal{A}$ for a linear wave is a section of Ω_t which moves with a fixed velocity 1 (i.e. dimensional velocity a_0) in a direction normal to the linear wavefront. For a nonlinear wavefront Ω_t the areas S_0 and S_1 used in the definitions of $\mathcal{A}$ are sections of Ω_t at t_0 and t, respectively moving with their velocity m. Here, Ω_t moves with a velocity m, which varies in space and time. We shall see from the next chapter onwards that there is a very interesting interaction of the wavefront velocity and geometry of Ω_t. This leads not only to longitudinal stretching of the rays but also a transverse deviation of the rays (or rotation of the wavefront) leading not only to a change of the mean curvature due to geometrical divergence or contraction of rays but also due to nonlinear change in the curvature Ω due to deformation of Ω_t. Due to these effects, the flux of energy density crossing a cross section of a ray tube is proportional to $(m-1)^2 e^{2(m-1)}$.

Let us recollect the ray coordinates introduced in the beginning of the Chap. 6. In 3-space dimensions (see Fig. 6.1) let $\boldsymbol{u}$ be the unit tangent vector along the surface coordinate ξ_1 (i.e., along the curve ($\xi_2 = const,\ t = const$)). Similarly, let $\boldsymbol{v}$ be the unit tangent vector along the surface coordinate ξ_2 (i.e. along the curve ($\xi_1 = const,\ t = const$)). Let χ be the angle[2] between the unit vectors $\boldsymbol{u}$ and $\boldsymbol{v}$, then the area of an elementary parallelogram with sides $g_1\boldsymbol{u}d\xi_1$ and $g_2\boldsymbol{v}d\xi_2$ on Ω_t is $g_1 d\xi_1 g_2 d\xi_2 \sin\chi$. Hence, it follows that $\mathcal{A} = g_1 g_2 \sin\chi$. In the Definition (7.1) of ray tube area, the tube in Fig. 7.1 is supposed not to collapse, i.e. it should remain

[2]The symbol χ is not to be confused with $\boldsymbol{\chi}$ used for the ray vector.

free from any singularity and hence $\mathcal{A} > 0$. Therefore, we choose $0 < \chi < \pi$. The Eq. (7.8) becomes

$$\left\{(m-1)^2 e^{2(m-1)} g_1 g_2 \sin\chi\right\}_t = 0, \quad 0 < \chi < \pi, \tag{7.10}$$

which is the final form of the energy transport equation along with a nonlinear ray in **conservation form**.

7.3 Transport Equations in 3-D SRT

For a shock front $\mathcal{S}(\boldsymbol{x}, t) = 0$ moving into an ambient gas at rest, the shock ray velocity $\boldsymbol{\chi}_s$ is ortogonal to the shock front (see first equation in (5.55)). Same argument as in the last section applies and the ray tube area $\mathcal{A}_s$ of the shock front is related to its mean curvature Ω_s by

$$\frac{1}{\mathcal{A}_s}\frac{\partial \mathcal{A}_s}{\partial \ell} = \langle \nabla, N \rangle \equiv -2\Omega_s. \tag{7.11}$$

Eliminating Ω_s between (5.56) and (7.11), and replacing d/dT by ∂_t in the ray coordinates and we get

$$\frac{2M}{M-1}M_t + \frac{\mathcal{A}_{st}}{\mathcal{A}_s} + 2M\mathcal{V} = 0, \tag{7.12}$$

where we have rearranged the terms in such a way that it follows the pattern of (7.7). This pattern is such that the first two terms give a combination $\{f'(h)/f(h)\}h_t + \mathcal{A}_{st}/\mathcal{A}_s$, where $h = M - 1$ and f is the function $f(h) := h^2 e^{2h}$. Hence, we get a conservation form

$$\{\mathcal{A}_s f(M-1)\}_t + 2\mathcal{A}_s M f(M-1)\mathcal{V} = 0. \tag{7.13}$$

In Sect. 4.4.2 we represented the shock front S_t in $\boldsymbol{x}$-space at any time t by $\mathcal{S}(\boldsymbol{x}, t) = 0$ with $t =$ const. We introduce a ray coordinate system (ξ_1, ξ_2, t) on the shock front S_t in usual way. Let $\boldsymbol{U}$ and $\boldsymbol{V}$ be respectively the unit tangent vectors along the curves ($\xi_2 =$ const, $t =$ const) and ($\xi_1 =$ const, $t =$ const) on Ω_t. We denote by G_1 and G_2 respectively, the associated metrics. On a given shock front at any time t we have the relations

$$\boldsymbol{X}_{\xi_1} = G_1 \boldsymbol{U}, \quad \boldsymbol{X}_{\xi_2} = G_2 \boldsymbol{V}, \tag{7.14}$$

where G_1 and G_2 are given by $G_1 = |\boldsymbol{X}_{\xi_1}|$ and $G_2 = |\boldsymbol{X}_{\xi_2}|$. Using an expression $\mathcal{A} = G_1 G_2 \sin\Psi$, where Ψ is the angle between the two unit vectors $\boldsymbol{U}$ and $\boldsymbol{V}$, for the ray tube area, we finally get from (7.12) a **balance equation** (equation in conservation form with additional source terms, in the absence of source terms we get conservation laws)

$$\left\{(M-1)^2 e^{2(M-1)} G_1 G_2 \sin\Psi\right\}_t + 2M(M-1)^2 e^{2(M-1)} G_1 G_2 \mathcal{V} \sin\Psi = 0. \quad (7.15)$$

This is the energy transport equation for SRT in conservation form.

Similarly, using the expression for Ω_s in (5.57), we obtain an equation which we rewrite as

$$\mathcal{V}_t + \frac{\mathcal{V}}{2\mathcal{A}}\mathcal{A}_{st} + \frac{\mathcal{V}}{2}\left(\frac{1}{M}-1\right)\frac{\mathcal{A}_{st}}{\mathcal{A}_s} + 2\mathcal{V}^2 = 0. \quad (7.16)$$

The first two terms of this equation are in form we wish and hence, we use (7.12) to replace the factor $\mathcal{A}_{st}/\mathcal{A}_s$ from the third term and write the resulting equation as

$$\left\{\ln\left(\mathcal{V}^2\mathcal{A}_s\right) + 2(M-1)\right\}_t + (M+1)\mathcal{V} = 0, \quad (7.17)$$

which gives a balance equation

$$\left\{e^{2(M-1)} G_1 G_2 \mathcal{V}^2 \sin\Psi\right\}_t + (M+1)e^{2(M-1)} G_1 G_2 \mathcal{V}^3 \sin\Psi = 0. \quad (7.18)$$

Remark 7.3.1 The second term in (7.12) and the second and third terms in (7.16) represent the geometrical effect of convergence or divergence of rays. The third term in (7.12) represents the effect of the interaction of nonlinear waves which overtake the shock from behind. The fourth term in (7.16) is the usual effect of genuine nonlinearity which governs the evolution of $\mathcal{V}$, also seen in the 1-D model $u_t + uu_x = 0$ (see the last term in (5.25), which for $i = 1$ becomes $-v_1^2$).

7.4 Energy Transport Equations in 2-D

In two space dimensions, the front Ω_t is a curve in (x, y)-plane. There is only one surface coordinate $\xi_1 = \xi$ and the ray coordinate plane is 2-D plane of (ξ, t). If we consider the geometry in 3-space dimensions, the ξ_2 coordinate is in the direction of x_3. The angle χ between ξ_1 and ξ_2 directions is $\pi/2$. From (7.10) and from (7.15) and (7.18) we get the following results.

Closure relation for 2-D WNLRT:

$$\left\{(m-1)^2 e^{2(m-1)} g\right\}_t = 0, \quad (7.19)$$

where g is the metric associated with the coordinate ξ.

Closure relations for 2-D SRT:

$$\left\{(M-1)^2 e^{2(M-1)} G\right\}_t + 2M(M-1)^2 e^{2(M-1)} G\mathcal{V} = 0 \tag{7.20}$$

and

$$\left\{e^{2(M-1)} G\mathcal{V}^2\right\}_t + (M+1)e^{2(M-1)} G\mathcal{V}^3 = 0, \tag{7.21}$$

where G is the metric associated with the coordinate ξ.

7.5 Kink and Triple Shock Interaction Point

One of the very important phenomena in gasdynamics is Mach reflection of a shock from a plane solid surface and has been discussed in a great detail in [14] in Sects. 128–136. In a Mach refection, the incident shock and the reflected shock meet at a point away from the surface and a Mach stem (also a shock) emerges from the point of intersection of these two and meets the reflecting surface. Thus in 2-D a Mach reflection consists of three shocks, namely the incident shock, the reflected shock and a Mach stem. There is also a slip plane (or slip line in 2-D), which is not a point of discussion here. The point where the three shocks meet is known as the **triple point** in two dimensions.

We shall see this triple point interaction in Sect. 9.1 when a caustic is resolved as in Fig. 9.2B. In this figure, we are looking at the geometry of a nonlinear wavefront. But the geometry of a shock front is qualitatively the same. The shock for $t > 0$ consists three parts, two wings on the two sides and a central disc. This shock has been produced by the sudden motion of a wedged shaped piston shown in the figure by $t = 0$. Imagine an incident shock falls on a plane surface (or a line $y = 0$ in this 2-D case), which has a Mach reflection, then, in this figure, the part of the shock above $y = 0$ consisting of the upper wing and the upper half of the central disc represents this phenomenon. The upper wing is the reflected shock and the upper half of the central disc is the Mach stem. The incident shock cannot be seen in this theory since we are following only the shock emerging from an initially given front in Fig. 9.2B.

Now suppose that the piston continues its motion after the sudden start, then it will send a succession of nonlinear wavefronts which will interact with the shock from behind (see Sect. 3.5). These nonlinear wavefront will also have a two wings and a central disc separated by a kink. The locus of the upper kinks in Fig. 9.2B as we move on the succession of the nonlinear wavefronts behind the shock drawn in the figure, will be the incident front.

The discussion in the above paragraph presents an open problem 'to reproduce the incident shock of a triple point interaction by the KCL theory'. One suggestion is to follow all nonlinear wavefronts produced at times $t > 0$ by a moving piston as in [8], Figs. 1 and 2, marked by 'Weakly nonlinear ray theory' and join the kinks on these nonlinear wavefronts behind the shock to get the incident shock. There are some more questions which have to be answered in this simulation of the incident shock.

Chapter 8
2-D KCL, WNLRT and SRT

In this chapter, we shall consider a particular case of KCL in two-space dimensions by taking $d = 2$ in the results in Chap. 6. We shall discuss a closed system of conservation laws by adding a transport equation for the amplitude of the moving curve, when this curve is a nonlinear wavefront and an additional transport equation for the gradient of pressure behind a shock, when this curve is a shock. We shall show that the KCL is well suited to study some beautiful physically realistic geometrical shapes of wavefronts. There is an extensive work on the use of 2D KCL to waves in a polytropic gas. Many of these results are available in [42]. Hence, we shall mention here some of the important old results briefly and continue with the results obtained after 2001 in Chap. 9.

We shall discuss mainly the isotropic evolution of nonlinear wavefronts and shock fronts. But, in the next section, we review kinematics of an anisotropic evolution of an arbitrary propagating curve.

8.1 Ray Coordinates (ξ, t) and 2-D KCL for Anisotropic Motion of Ω_t

We shall review the general theory of 2-D KCL for an anisotropic motion of Ω_t, which has been discussed in detail in [42] in Sects. 3.3.2 and 3.3.3. We take the ray velocity $\boldsymbol{\chi} = (\chi_1, \chi_2)$ to be not necessarily orthogonal to Ω_t and note the relation (4.3) between $\boldsymbol{\chi}$ and the normal velocity C. We write the equation of Ω_t in the form

$$\Omega_t : x = x(\xi, t), \quad y = y(\xi, t), \tag{8.1}$$

where constant values of t give the positions of the propagating curve Ω_t at different times and $\xi =$ constant represents a ray. Let the unit normal of Ω_t be $\boldsymbol{n} = (n_1, n_2)$. Velocity C of Ω_t and the component T of the ray velocity in the direction of the tangent to the curve Ω_t are related to χ_1 and χ_2 by

P. Prasad, *Propagation of Multidimensional Nonlinear Waves and Kinematical Conservation Laws*, Infosys Science Foundation Series,
https://doi.org/10.1007/978-981-10-7581-0_8

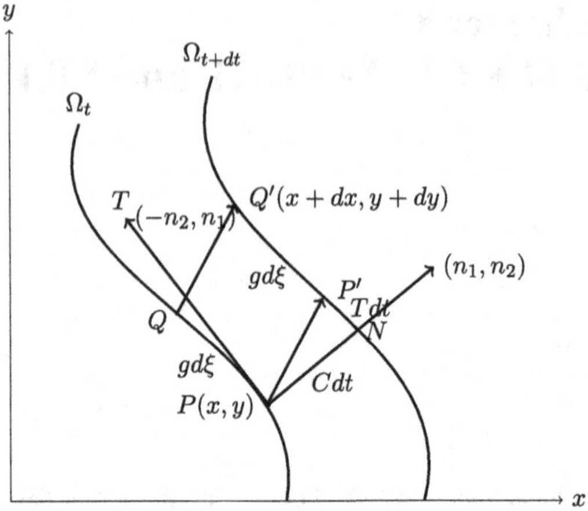

Fig. 8.1 P' and Q' on Ω_{t+dt} are positions of P and Q, respectively, as we move on two rays. $PQ = gd\xi = P'Q'$, $PN = Cdt$ and $NP' = Tdt$

$$C = n_1\chi_1 + n_2\chi_2,\ T = -n_2\chi_1 + n_1\chi_2;\quad \chi_1 = Cn_1 - Tn_2,\ \chi_2 = Cn_1 + Tn_2. \tag{8.2}$$

Derivation of 2-D KCL: (ξ, t) is a ray coordinate system and let g be the metric associated with ξ. Consider the curves Ω_t and Ω_{t+dt} (see Fig. 8.1). Let P and Q be two points on Ω_t at a distance $gd\xi$ and let these points reach points P' and Q' on Ω_{t+dt}, while moving along rays. $PN = Cdt$ is the normal displacement of Ω_t in time dt. If the coordinates of Q' are $(x+dx, y+dy)$, then (dx, dy) is a displacement in the (x, y) plane corresponding to a displacement $(d\xi, dt)$ in the ray coordinate plane, so that

$$\begin{aligned} dx &= (Cdt)n_1 - (gd\xi + Tdt)n_2 \\ dy &= (Cdt)n_2 + (gd\xi + Tdt)n_1 \end{aligned} \tag{8.3}$$

The above relation gives the Jacobian matrix of the transformation from (ξ, t)-plane to (x, y)-plane[1]

$$\begin{pmatrix} x_\xi & x_t \\ y_\xi & y_t \end{pmatrix} = \begin{pmatrix} -gn_2 & Cn_1 - Tn_2 = \chi_1 \\ gn_1 & Cn_2 + Tn_1 = \chi_2 \end{pmatrix}. \tag{8.4}$$

The determinant of the Jacobian matrix, is $-gC$, which shows that the transformation between (ξ, t) and (x, y) is non-singular as long as g and C are non-zero and finite. We have made some comments on the singularities of the transformation and special type of singularities **kinks** in (x, y)-plane in Sect. 6.3 (see also Sect. 3.3.3 of [42] for details) (Fig. 8.1).

In ray coordinates ξ and t, the partial derivative $\frac{\partial}{\partial t}$ represents the time-rate of change along a ray and equals to $\frac{d}{dt} = \frac{\partial}{\partial t} + \chi_1\frac{\partial}{\partial x_1} + \chi_2\frac{\partial}{\partial x_2}$ in (x_1, x_2, t)-space. It is simple to get the expression for $\frac{\partial}{\partial t}$ and $\frac{\partial}{\partial \xi}$ in (ξ, t)-plane, $\left(\frac{\partial}{\partial t}\right)_{(\xi,t)plane} = x_t\frac{\partial}{\partial x} + y_t\frac{\partial}{\partial y}$

[1]Care is required when we are talking of (ξ, t)-plane, which is the same as (x, y)-plane and not to be confused with 3-D (x, y, t)-space.

and $\left(\frac{\partial}{\partial \xi}\right)_{(\xi,t)plane} = x_\xi \frac{\partial}{\partial x} + y_\xi \frac{\partial}{\partial y}$ and use chain rule from (8.2). There is no need to keep the subscript $(\xi, t)plane$, and we have the expressions

$$\frac{\partial}{\partial t} = \left(\chi_1 \frac{\partial}{\partial x} + \chi_2 \frac{\partial}{\partial y}\right) \quad and \quad \frac{\partial}{g\partial \xi} = -n_2 \frac{\partial}{\partial x} + n_1 \frac{\partial}{\partial y}. \tag{8.5}$$

When the evolution of Ω_t is smooth, we use (8.4) to equate $x_{\xi t} = x_{t\xi}$ and $y_{\xi t} = y_{t\xi}$ and derive a pair of relations in the conservation form, i.e. the 2-D KCL:

$$(g \sin\theta)_t + (C \cos\theta - T \sin\theta)_\xi = 0, \tag{8.6}$$

$$(g \cos\theta)_t - (C \sin\theta + T \cos\theta)_\xi = 0, \tag{8.7}$$

where θ is the angle which normal to Ω_t makes with the x-axis, i.e. $\boldsymbol{n} = (\cos\theta, \sin\theta)$.

In order to satisfy a conservation law, the functions θ, C and T need not be smooth but when they are smooth we deduce from (8.6) and (8.7) the following partial differential equations

$$g_t = C\theta_\xi + T_\xi, \quad \theta_t = -\frac{1}{g}C_\xi + \frac{1}{g}T\theta_\xi \tag{8.8}$$

as kinematical relations for any propagating curve Ω_t. Equations (8.8) or their conservation forms (8.6) and (8.7) represent a system of two equations involving four quantities g, θ, C and T. Thus, as we noticed in the general case of d-D KCL, this system is under-determined.

We quote here a particular case of the Theorem 6.2.1 (see also [42]):

Theorem 8.1.1 *For smooth curve Ω_t the KCL system (8.6) and (8.7) is equivalent to the ray equations (4.6) and (4.7), which we write explicitly for 2-D:*

$$\frac{dx}{dt} = \chi_1, \quad \frac{dy}{dt} = \chi_2, \quad \frac{d\theta}{dt} = -\frac{1}{g}(n_1 \frac{\partial \chi_1}{\partial \xi} + n_2 \frac{\partial \chi_2}{\partial \xi}) \equiv -\frac{1}{g}C_\xi + \frac{1}{g}T\theta_\xi. \tag{8.9}$$

Jump relation and conservation of distance: The jump relations or R-H conditions obtained from the conservation laws (8.6) and (8.7) are

$$-S[g \sin\theta] + [C \cos\theta - T \sin\theta] = 0, \quad S[g \cos\theta] + [C \sin\theta + T \cos\theta] = 0 \tag{8.10}$$

where S is the velocity of the shock along ξ-axis. Eliminating S from these, we get the following **Hugoniot curve**:

$$\cos(\theta_\ell - \theta_r) = \frac{C_\ell g_\ell + C_r g_r}{C_\ell g_r + C_r g_\ell} + \frac{g_r T_\ell - g_\ell T_r}{g_r C_\ell + g_\ell C_r} \sin(\theta_\ell - \theta_r), \tag{8.11}$$

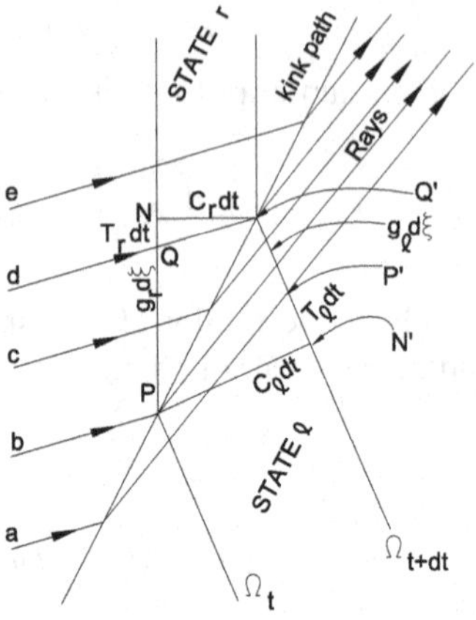

Fig. 8.2 P' and Q' are positions on Ω_{t+dt} of P and Q, respectively, on two rays at distance $gd\xi$ on Ω_t

which is one relation between quantities g_ℓ, C_ℓ, , θ_ℓ and T_ℓ, say on the negative side, in terms of quantities on the positive side of the shock.

For Ω_t moving in anisotropic medium, Fig. 8.2 represents a kink phenomenon in the (x, y)-plane corresponding to a forward facing shock which moves in ξ increasing direction (larger root of (8.13)). The kink in Fig. 6.2 corresponds to a backward facing shock (smaller root of (8.13) with $T = 0$). The kink at P on Ω_t occupies a position Q' on Ω_{t+dt}. The kink path PQ' separates two states represented by subscripts r and ℓ. PN' and $Q'N$ are the normals from P and Q' to the negative side of Ω_{t+dt} and positive side of Ω_t, respectively. $N'P' = T_\ell dt$ is the displacement of a ray along Ω_{t+dt} due to the tangential velocity T_ℓ and $g_\ell d\xi$ is the distance along Ω_{t+dt} between P' and Q$'$. Corresponding quantities $g_r d\xi$ and $T_r dt$ are PQ and QN, respectively. Using the Pythagoras theorem for the two right-angled triangles $PQ'N$ and $PQ'N'$, we get

$$(g_r\, d\xi + T_r dt)^2 + (C_r dt)^2 = (g_\ell d\xi + T_\ell dt)^2 + (C_\ell dt)^2 \tag{8.12}$$

which shows that the kink velocity $S = d\xi_p/dt$ satisfies the quadratic equation

$$(g_\ell^2 - g_r^2)S^2 + 2(g_\ell T_\ell - g_r T_r)S + (C_\ell^2 - C_r^2) \ + \ (T_\ell^2 - T_r^2) = 0. \tag{8.13}$$

This equation has real roots for S if

$$(g_\ell T_r - g_r T_\ell)^2 + (C_r^2 - C_\ell^2)(g_\ell^2 - g_r^2) > 0. \tag{8.14}$$

Equation (8.13) is a beautiful result obtained purely from geometry and can also be derived by eliminating θ from the jump relation (8.10) of KCL (8.6) and (8.7). This only verifies the statement in the Theorem 6.3.2: *the KCL are physically realistic.*

2-D KCL for isotropic evolution of Ω_t: In this case $T = 0$ and we denote the normal velocity by m, i.e. $C = m$, as in the Sect. 6.1. Then $\chi_1 = m\cos\theta$ and $\chi_2 = m\sin\theta$. The ray equations (8.9) for isotropic evolution of Ω_t give

$$\frac{dx}{dt} = m\cos\theta, \quad \frac{dy}{dt} = m\sin\theta; \qquad \frac{d\theta}{dt} = -(-\sin\theta\frac{\partial}{\partial x} + \cos\theta\frac{\partial}{\partial y})m,$$

which in ray coordinates become

$$x_t = m\cos\theta, \quad y_t = m\sin\theta; \qquad \theta_t = -\frac{1}{g}m_\xi. \tag{8.15}$$

The metric g is given by $g^2 = x_\xi^2 + y_\xi^2$. Differentiating it with respect to t and using the first two equations in (8.15), we can derive $g_t - m\theta_\xi = 0$. Thus, Eq. (8.15) lead to a system of two equations for m, θ *and* g in ray coordinates (ξ, t)

$$\theta_t + \frac{1}{g}m_\xi = 0, \quad g_t - m\theta_\xi = 0. \tag{8.16}$$

In this case, the KCL (8.6) reduces to

$$(g\sin\theta)_t + (m\cos\theta)_\xi = 0, \quad (g\cos\theta)_t - (m\sin\theta)_\xi = 0. \tag{8.17}$$

Note that (8.16) is a differential form of the the KCL (8.17) but have been derived from the ray equations, i.e. the bicharacteristic equations of the wave equation $u_{tt} - m^2\left(u_{xx} + u_{yy}\right) = 0$, where m need not be constant (see (6.2) and (6.3)). Thus, we note that the differential form (8.16) of the KCL (8.17) can also be obtained from the ray equations (8.15). *This statement is a beautiful example equivalence of a result in PDE and a result in geometry.* It is also related to the Theorem 6.2.1. We have already remarked in historical results in the beginning of the Chap. 6 that Eq. (8.16) were derived before we derived the 2-D KCL.

8.2 2-D WNLRT and SRT for Polytropic Gas: An Isotropic Media

Non-dimensional form of the weakly nonlinear ray theory (WNLRT) and the shock ray theory (SRT) have been written down in the Sect. 5.4 and their conservation forms in Sects. 7.4 and 8.4. In this section we collect all these scattered equations together. First, we write the conservation forms and then their differential forms.
Equations of WNLRT: We first note that the Theorem 8.1.1 implies that the ray equations (8.15) are equivalent to the KCL (8.17) for a smooth function satisfying (8.17). But KCL equations are more basic. The conservation forms of equations of WNLRT in 2-D are simply (8.17) and that of the energy transport equation is (7.20).

These equations are

$$(g\sin\theta)_t + (m\cos\theta)_\xi = 0,\quad (g\cos\theta)_t - (m\sin\theta)_\xi = 0, \tag{8.18}$$

and

$$\left\{(m-1)^2 e^{2(m-1)} g\right\}_t = 0. \tag{8.19}$$

The differential forms of (8.18) and (8.19) are

$$\theta_t + \frac{1}{g} m_\xi = 0,\quad g_t - m\theta_\xi = 0, \tag{8.20}$$

and

$$m_t + \frac{m-1}{2g}\theta_\xi = 0. \tag{8.21}$$

(8.21) also follows from (5.53) after using $\Omega = \frac{1}{2}\left(\sin\theta\frac{\partial\theta}{\partial x} - \cos\theta\frac{\partial\theta}{\partial y}\right)$ from (5.6) and $\frac{\partial\theta}{\partial\xi} = -g\left(\sin\theta\frac{\partial\theta}{\partial x} - \cos\theta\frac{\partial\theta}{\partial y}\right)$ from (8.5).

Now we have a complete set of equations for three unknowns m, θ and g. **Note that WNLRT and SRT below are physically realistic only for** $0 < m - 1 << 1$ **and** $0 < M - 1 << 1$.

Equations of SRT: The conservation form of SRT are KCL (8.17) (with $m,\ \theta$ and g replaced, respectively, by $M,\ \Theta$ and G) and the closure relations are (7.20) and (7.21) (see also [8]). Here M is the shock Mach number, G the metric associated with the coordinate ξ and Θ is the angle which shock normal $\boldsymbol{N}$ makes with the x-axis. Thus, the conservation forms of SRT are

$$(G\sin\Theta)_t + (M\cos\Theta)_\xi = 0,\quad (G\cos\Theta)_t - (M\sin\Theta)_\xi = 0, \tag{8.22}$$

$$\left\{(M-1)^2 e^{2(M-1)} G\right\}_t + 2M(M-1)^2 e^{2(M-1)} G\mathcal{V} = 0 \tag{8.23}$$

and

$$\left\{e^{2(M-1)} G\mathcal{V}^2\right\}_t + (M+1)e^{2(M-1)} G\mathcal{V}^3 = 0, \tag{8.24}$$

where $\mathcal{V}$, defined in (5.50), is a measure of the gradient in the normal direction of the pressure or density just behind the shock and takes into account of the effect of interactions of the nonlinear waves which catch-up with the shock from behind.

From (8.16), (5.56) and (5.57), the differential forms of SRT are

$$\Theta_t + \frac{1}{G} M_\xi = 0,\quad G_t - M\Theta_\xi = 0, \tag{8.25}$$

$$M_t + \frac{1}{2G}(M-1)\Theta_\xi = -(M-1)\mathcal{V}, \tag{8.26}$$

and

$$\mathcal{V}_t + \frac{\mathcal{V}}{2G}\Theta_\xi = -2\mathcal{V}^2. \tag{8.27}$$

We have a complete set of equations for four unknowns M, Θ, G and $\mathcal{V}$.

8.3 Mapping from (ξ, t)-plane to Physical (x, y)-plane

Let us describe the mapping from (ξ, t)-plane to the (x, y)-plane when a solution of the equations of WNLRT in (ξ, t)-plane has been obtained. Similar procedure is valid for a solution of the equations of SRT. Assume that we have a solution of the equations of (8.18) and (8.19) in the form

$$m = \tilde{m}(\xi, t), \quad \theta = \tilde{\theta}(\xi, t), \quad g = \tilde{g}(\xi, t). \tag{8.28}$$

We map this solution onto (x, y)-plane, in which the nonlinear wavefront Ω_t propagates. The aim is to find the successive positions of Ω_t and the rays associated with it. Let the initial position of the wavefront be given by

$$\Omega_0 : \quad x = x_0(\xi), \quad y = y_0(\xi). \tag{8.29}$$

Method of calculation of Ω_t from solution in (ξ, t)-plane: Apart from the first two equations $x_t = m\cos\theta$, $y_t = m\sin\theta$ in (8.15), we also have two more equations in (8.4)

$$x_\xi = -g\sin\theta, \quad y_\xi = g\cos\theta. \tag{8.30}$$

Let us take a point $P_0 : (x(\xi_0, 0), y(\xi_0, 0))$ on Ω_0. Given the solution (8.28), we integrate the first two of ray equations (8.15)

$$\frac{dx}{dt} = \tilde{m}(\xi, t)\cos\tilde{\theta}(\xi, t) \quad \frac{dy}{dt} = \tilde{m}(\xi, t)\sin\tilde{\theta}(\xi, t) \tag{8.31}$$

along a ray, say $\xi = \xi_0 =$ a constant, starting from P_0 reaching a point $(x(\xi_0, t), y(\xi_0, t))$, see Fig. 8.3. Then we integrate (8.30), i.e.,

$$x_\xi = -\tilde{g}(\xi, t)\sin\tilde{\theta}(\xi, t), \quad y_\xi = \tilde{g}(\xi, t)\cos\tilde{\theta}(\xi, t) \tag{8.32}$$

along the curve Ω_t starting from the point $(x(\xi_0, t), y(\xi_0, t))$ and reach any point $P : (x(\xi, t), y(\xi, t))$. This gives the nonlinear wavefront Ω_t and also the mapping

$$x = x(\xi, t), \quad y = y(\xi, t) \tag{8.33}$$

from the ray coordinate plane to the physical plane.

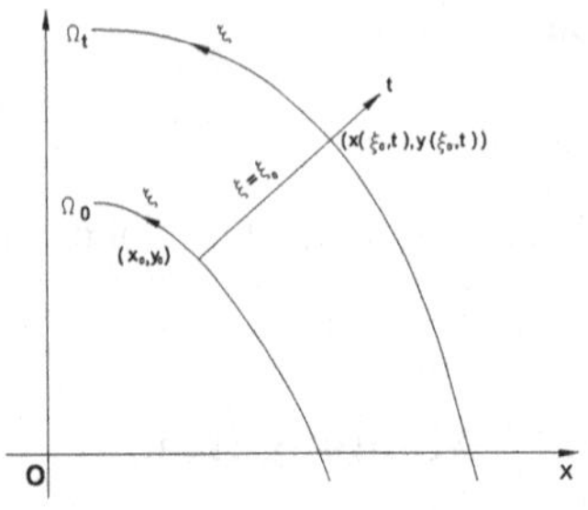

Fig. 8.3 Calculation of Ω_t from the solution in (ξ, t)-plane

Remark 8.3.1 We remind a simple result: a line parallel to the ξ-axis is mapped onto a wavefront and a line parallel to t-axis is onto a ray.

Therefore, use of ray coordinates makes the problem of calculation of successive positions of Ω_t and tracing rays very simple.

8.4 Eigenvalues and Elementary Wave Solutions

The eigenvalues of the WNLRT Eqs. (8.20) and (8.21) are

$$c_1 = -\sqrt{\frac{m-1}{2g^2}}, \quad c_2 = 0, \quad c_3 = \sqrt{\frac{m-1}{2g^2}}. \tag{8.34}$$

The characteristic fields of c_1 and c_2 are genuinely nonlinear and that of c_1 is linearly degenerate.

The independent Riemann invariants (see Definition 4.2.4) corresponding to the ith characteristic fields are denoted by $(\pi_1^{(i)}, \pi_2^{(i)})$ for $i = 1, 2, 3$ and are given by

$$\pi_1^{(1)} = \theta + \sqrt{8(m-1)}, \ \pi_2^{(1)} = g(m-1)^2 e^{2(m-1)}; \qquad \pi_1^{(2)} = m, \ \pi_2^{(2)} = \theta;$$
$$\pi_1^{(3)} = \theta - \sqrt{8(m-1)}, \ \pi_2^{(3)} = g(m-1)^2 e^{2(m-1)}. \tag{8.35}$$

Jump relations for WLNLRT: The R-H jump conditions from Eqs. (8.18) and (8.19), joining two states $(m, \theta, g)_\ell$ and $(m, \theta, g)_r$ give

$$-S[g \sin\theta] + [m\cos\theta] = 0, \quad S[g\cos\theta] + [m\sin\theta] = 0 \tag{8.36}$$

and

$$S(g_l(m_\ell - 1)^2 e^{2(m_\ell - 1)} - g_r(m_r - 1)^2 e^{2(m_r - 1)}) = 0. \tag{8.37}$$

If $S = 0$, the Eq. (8.37) is automatically satisfied. Equation (8.36) gives $m_r = m_\ell$ and $\theta_r = \theta_\ell$. Both equations are satisfied by $g_\ell \neq g_r$. This discontinuity appearing in the degenerate characteristic field of $c_2 = 0$ is a **contact discontinuity** C, which we shall take up later in this section. For a detailed discussion on an elementary wave which is a contact discontinuity, see [51] (pages 333 and 334).

For a discontinuity in first and third characteristic field $S \neq 0$. Eliminating S from (8.36) we get (see also (8.11))

$$\cos(\theta_r - \theta_l) = \frac{m_l g_l + m_r g_r}{m_l g_r + m_r g_l} \tag{8.38}$$

and eliminating θ_r and θ_ℓ from (8.36) we get[2]

$$S^2 = \frac{m_l^2 - m_r^2}{g_r^2 - g_l^2}. \tag{8.39}$$

There is another expression[3] for the shock velocity S

$$S = \frac{(m_r^2 - m_l^2)}{(m_l g_r + m_r g_l)\sin(\theta_r - \theta_l)}. \tag{8.40}$$

In [42] the expression (2.8) for S is to be corrected as shown above.

Equation (8.40) is used to get the sign of $\theta_r - \theta_\ell$ and hence direction of the kink line in (x, y)-plane, see below (8.48) and Fig. 8.6.

Definition 8.4.1 **Elementary wave solutions** of the system of conservation (4.78) are solutions of the form $\boldsymbol{u}(x, t) = \boldsymbol{u}(\frac{x - x_0}{t - t_0})$. These are centred rarefaction wave solutions with the centre at (x_0, t_0) in (x, t)-plane and shock waves passing through this point.

Elementary wave solutions appearing in a Riemann problem: In Sect. 3.3, we discussed Riemann problem for the single conservation law (3.16) which led to two types of *elementary wave solutions* (3.19) and (3.20). The initial data for a Riemann problem for the WNLRT system (8.18) and (8.19) is

$$(m, \theta, g)(\xi, 0) = \begin{cases} (m_0, \theta_0, g_0)_\ell & , \xi \leq 0 \\ (m_0, \theta_0, g_0)_r & , \xi > 0 \end{cases}, \tag{8.41}$$

where $(m_0, \theta_0, g_0)_\ell$ and $(m_0, \theta_0, g_0)_r$ are constant states with both $m_{0\ell}$ and $m_{0r} > 1$.

In general, the solution of this problem in (ξ, t)-plane, $t > 0$, consists of four constant state regions separated by **elementary waves** [51] of the three families of

---

[2]Derivation is simple. In the jump relations, we collect all terms containing θ_ℓ on the left-hand side and those containing θ_r on the right-hand side. Then, we square and add the two jump relations to get (8.39).

[3]Derivation requires multiplying the first jump relation in (8.36) by $\cos\theta_\ell$, the second by $\sin\theta_\ell$ and adding. Then we get $Sg_r \sin(\theta_\ell - \theta_r) + m_r \cos(\theta_\ell - \theta_r) - m_\ell = 0$, in which we use (8.38) to get (8.40).

characteristics. The elementary wave of the first family separates the constant state region with state $(m_0, \theta_0, g_0)_\ell$ on the left and a constant state region with a new state $(m, \theta, g)_{\ell i}$ on the right. The elementary wave of the third family separates constant state regions with a new state $(m, \theta, g)_{ri}$ on the left and the state $(m_0, \theta_0, g_0)_r$ on the right. The two constant state regions with constant values $(m, \theta, g)_{\ell i}$ and $(m, \theta, g)_{ri}$ are separated by an elementary wave of the second family.

Since the first and third families of characteristics fields are genuinely nonlinear, the elementary waves of these two families are either shocks or a simple waves. The intermediate elementary wave of the second family is a **contact discontinuity** $\mathcal{C}$ along the t-axis and is bounded by constant state solutions $(m, \theta, g)_{\ell i}$ on the left and $(m, \theta, g)_{ri}$ on the right. Since $S = c_2 = 0$ for $\mathcal{C}$, as mentioned just below (8.36) and (8.37), $(m, \theta)_{\ell i} = (m, \theta)_{ri}$ and $g_{\ell i} \neq g_{ri}$.

Contact discontinuity and the energy equation for an initial value problem in (ξ, t)-plane: Consider an initial value problem for the system (8.18) and (8.19)

$$(m, \theta, g) = (m_0(\xi), \theta_0(\xi), g_0(\xi)) \tag{8.42}$$

and assume that the solution has an isolated contact discontinuity $\mathcal{C}$, say at $\xi = 0$ for $0 < t < t_c$.

In a domain in (ξ, t)-plane, where the solution is smooth, the energy equation (8.19) can be integrated to give

$$(m-1)^2 e^{2(m-1)} g = (m_0 - 1)^2 e^{2(m_0-1)} g_0 =: f(\xi). \tag{8.43}$$

Apart from the contact discontinuity $\mathcal{C}$, there may be shocks of the first and third family with a shock velocity $S \neq 0$ for $0 < t < t_c$. When a line $\xi = \xi_0 \neq 0$ meets a shock, the shock jump relation (8.37) gives $(m-1)^2 e^{2(m-1)} g = (m_0 - 1)^2 e^{2(m_0-1)} g_0 =: f(\xi)$. Thus (8.43) is valid everywhere for $0 < t < t_c$ except on $\xi = 0$. Across C at $\xi = 0$ we have $g_{0\ell} \neq g_{0r}$.

Now we define ξ' such that

$$\xi' = \int^{\xi} f(\xi) d\xi. \tag{8.44}$$

Since for an element ds of arc length of Ω_t, we have expressions $ds = g d\xi = g' d\xi'$, a choice of coordinate ξ' according to (8.44) implies $g = g' f(\xi)$. The relation (8.43) becomes $(m-1)^2 e^{2(m-1)} g' = 1$. We have proved a very useful theorem:

Theorem 8.4.2 *In a solution of an initial value problem for the conservation laws of WNLRT containing a number of isolated contact discontinuities, it is possible to make a suitable choice of ξ, so that we can take the energy transport equation in the form:*

$$g = (m-1)^{-2} e^{-2(m-1)}. \tag{8.45}$$

Note: In next three sections, namely Sects. 8.5, 8.6 and 8.7, we shall assume that transformation (8.44) has been made and we shall always use the relation (8.45) instead of the conservation form of the energy transport equation (8.19). Therefore, no contact discontinuity will appear in any solution we consider.

8.5 Simple Waves and Shocks of WNLRT Conservation Laws

These solutions of (8.18) and (8.45) have been discussed briefly in [42] and in great detail in [7] with a closure relation more general than (8.45). We mention here only some salient features. We wish to join a state (m_ℓ, θ_ℓ) on the left in (ξ, t)-plane by a simple wave or a shock to a state (m_r, θ_r) on the right.
A choice of the angle θ_ℓ: By rotation of the coordinate axes in (x, y)-plane we can always choose $\theta_\ell = 0$.

1-R wave: is a symbol to denote a simple wave (see Definition 4.2.6 in Sect. 4.2.2) in c_1 characteristic family. Note that $c_1 < 0$. In this wave, the Riemann invariants $\pi_1^{(1)} = \theta + \sqrt{8(m-1)}$ and $\pi_2^{(1)} = g(m-1)^2 e^{2(m-1)}$ are constants (we have chosen ξ so that $\pi_2^{(1)} = g(m-1)^2 e^{2(m-1)} = 1$), and hence

$$\theta + \sqrt{(8(m-1))} = \sqrt{(8(m_\ell - 1))}. \tag{8.46}$$

This curve in (θ, m)-plane has been denoted by $R_1(m_\ell)$.

However, we are interested in a **centred simple or rarefaction wave** (see Definition 4.2.11 and text after (4.30)) of the first family in which the slope $c_1^{-1} = -\left(\sqrt{\frac{m-1}{2g^2}}\right)^{-1} = -\frac{\sqrt{2}}{(m-1)^{5/2}e^{2(m-1)}}$ of the characteristics of c_1 family continuously increases in (ξ, t)-plane as we move in it from left to right. Hence, it follows that m in a centered 1-R wave continuously decreases as ξ increases on a line $t = constant$, and m in the interior of centred 1-R wave and in the state on the right of it satisfy $m_r < m < m_\ell$. We denote by $R_1^-(m_\ell, \theta_\ell = 0)$ the curve in (m, θ)-plane represented by

$$\boldsymbol{R_1^-(m_\ell)} =: \ \theta + \sqrt{(8(m-1))} = \sqrt{(8(m_\ell - 1))}, \ \ 1 < m < m_\ell. \tag{8.47}$$

All states (m_r, θ_r), which can be joined to the state $(m_\ell, \theta_\ell = 0)$ by a centred 1-R wave lie on $R_1^-(m_\ell)$ on the left the point P_ℓ $(m_\ell, 0)$ in (m, θ)-plane.

The curve $R_1^-(m_\ell)$ (and other related curves $R_3^+(m_\ell)$, $S_1^+(m_\ell)$ and $S_3^-(m_\ell)$ defined below) have been drawn in Fig. 8.4. In this figure the point P_ℓ is the point from where these curves originate.

1-S shock: is a shock wave in c_1 characteristic field and its Hugoniot curve is given by (8.38) with $\theta_\ell = 0$. The Lax entropy condition in the form $c_1(u_r) < S < c_1(u_\ell)$ implies $\frac{1}{\sqrt{2}}(m_\ell - 1)^{\frac{5}{2}} e^{2(m_\ell - 1)} < \frac{1}{\sqrt{2}}(m_r - 1)^{\frac{5}{2}} e^{2(m_r - 1)}$, i.e., $m_\ell < m_r$. Now (8.40), since $S < 0$, gives

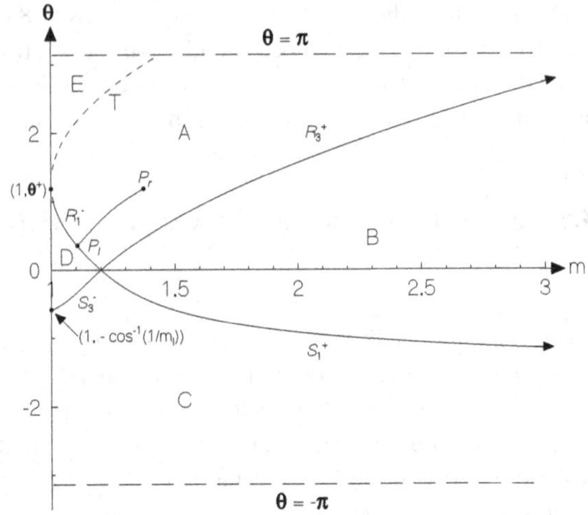

Fig. 8.4 Points in (m, θ)-plane which can be joined to a state $(m_\ell, \theta_\ell = 0)$ through centred rarefaction curves $R_1^-(m_\ell)$ and $R_3^+(m_\ell)$; and Hugoniot curves $S_1^+(m_\ell)$ and $S_3^-(m_\ell)$. Point P_i has not been shown. This figure has been reproduced from [7] - Fig. 4, with permission from the Publisher, Oxford University Press

$$-\frac{\pi}{2} < \theta_r - \theta_\ell < 0, \tag{8.48}$$

and (8.38) gives $S_1^+(m_\ell)$ curve (note we take $\theta_\ell = 0$)

$$\boldsymbol{S_1^+(m_\ell)}: \ \theta = -\cos^{-1}\left(\frac{m_\ell g_\ell + mg}{m_\ell g + m g_\ell}\right), \quad m_\ell < m, \tag{8.49}$$

where we take only the positive determination of the $\cos^{-1}$ function.

Therefore, all the points (m_r, θ_r) which can be joined to the state $(m_\ell, \theta_\ell = 0)$ on the left by 1-S shock lie on the curve $S_1^+(m_\ell)$ on the right of the point $(m_\ell, 0)$ in (m, θ)-plane. Note that in (8.49) $\lim_{m\to\infty} \theta = -\frac{\pi}{2}$.

3-R wave and 3-S shock: As above, we define 3-R simple wave in which the Riemann invariants $\pi_1^{(3)} = \theta - \sqrt{8(m-1)}$ and $\pi_2^{(3)} = g(m-1)^2 e^{2(m-1)}$ are constants, $\pi_2^{(3)} = g(m-1)^2 e^{2(m-1)} = 1$. The $\boldsymbol{R_3^+(m_\ell)}$ is the set of points in (m, θ)-plane which can be joined to the state $(m_\ell, \theta_\ell = 0)$ on the left by a **3-R centred** simple wave:

$$\boldsymbol{R_3^+(m_\ell)} =: \ \theta - \sqrt{(8(m-1))} = -\sqrt{(8(m_\ell - 1))}, \ 1 < m_\ell < m. \tag{8.50}$$

Similarly, we can define 3-S shock for which

$$\boldsymbol{S_3^-(m_\ell)} =: \ \theta = -\cos^{-1}\left(\frac{m_l g_l + mg}{m_l g + m g_l}\right), \quad m < m_\ell, \tag{8.51}$$

which is the set of points in (m, θ)-plane which can be joined to the state $(m_\ell, \theta_\ell = 0)$ on the left through a 3-S shock.

Elementary shapes $\mathcal{R}_1$, $\mathcal{R}_3$ and kinks $\mathcal{K}_1$ and $\mathcal{K}_3$: Consider centred simple or rarefaction waves 1-R, 3-R; and shocks 1-S and 3-S solutions in (ξ, t)-plane with

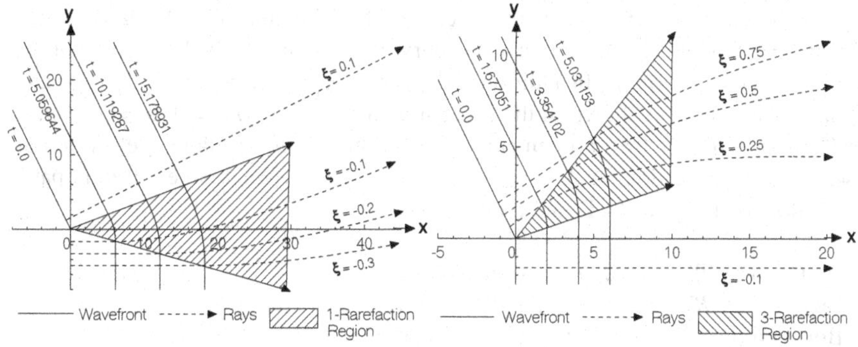

Fig. 8.5 8.5A: $\mathcal{R}_1$ elementary shape with $m_\ell = 1.2,\ m_r = 1.08$. Rays enter the $\mathcal{R}_1$ shape from below and $\mathcal{R}_1$ moves down on the nonlinear wavefront Ω_t. 8.5B: $\mathcal{R}_3$ elementary shape with $m_\ell = 1.2,\ m_r = 1.4$. Rays enter the $\mathcal{R}_3$ shape from above and $\mathcal{R}_3$ moves up on the nonlinear wavefront Ω_t. These figures have been reproduced from [7]- Figs. 6 and 7, with permission from the Publisher, Oxford University Press

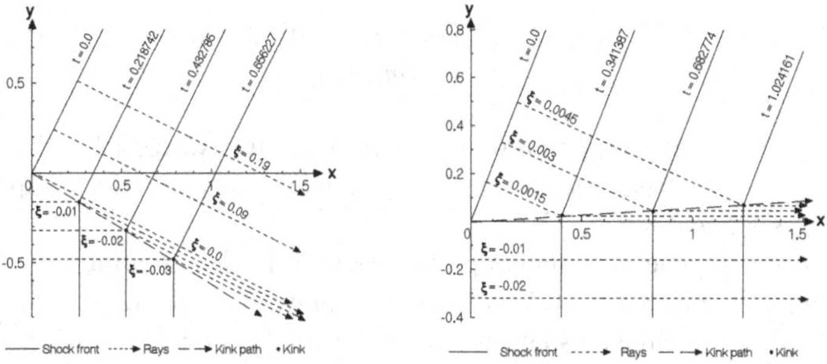

Fig. 8.6 8.6A: Kink $\mathcal{K}_1$ with $m_\ell = 1.2,\ m_r = 1.4$. Rays intersect the kink line from below and $\mathcal{K}_1$ moves down on the nonlinear wavefront Ω_t. 8.6B: Kink $\mathcal{K}_3$ with $m_\ell = 1.2,\ m_r = 1.08$. Rays intersect the kink line from above and $\mathcal{K}_3$ moves up on the nonlinear wavefront Ω_t. These figures have been reproduced from [7] - Figs. 8 and 9, with permission from the Publisher, Oxford University Press

center at ($\xi = 0, t = 0$). Images of these rarefaction wave solutions 1-R and 3-R in (x, y)-plane give smooth curve parts of Ω_t, which are elementary shapes $\mathcal{R}_1$ and $\mathcal{R}_3$. These elementary shapes are seen in Fig. 8.5. Similarly, images of 1-S and 3-S solutions give elementary shapes $\mathcal{K}_1$ and $\mathcal{K}_3$ on Ω_t, which are kinks. We see them in Fig. 8.6.

8.6 Solution of the Riemann Problem

The results in this section also have been discussed briefly in [42] and in great detail in [7]. We first note from (8.45) that $g \to \infty$ as $m \to 1+$. In the

Fig. 8.4 the curve $R_1^-(m_\ell)$ meets the line $m = 1$ at the point $(1, \theta^+((m_\ell))$, where $\theta^+((m_\ell) = \sqrt{(8(m_\ell - 1))}$. Similarly, the curve $S_3^-(m_\ell)$ meets the line $m = 1$ at the point $(1, -\cos^{-1}(1/(m_\ell))$. In addition to the curves $R_1^-(m_\ell)$, $R_3^+(m_\ell)$, $S_1^+(m_\ell)$ and $S_3^-(m_\ell)$, we have also shown in the Fig. 8.4 a part T of $R_3^+(m = 1, \theta = \theta^+(m_\ell))$ for $\theta^+(m_\ell) < \theta < \pi$ by a broken curve. The points in (m, θ)-plane relevant to our discussion lie in the domain $1 < m < \infty$, $-\pi < \theta < \pi$. We denote different parts of this domain by A, B, C, D and E as follows:

A: Bounded by $R_1^-(m_\ell), T, \theta = \pi$ and R_3^+,
B: Bounded by $R_3^+(m_\ell)$, $S_1^+(m_\ell)$ and $m = \infty$,
C: Bounded by $S_3^-(m_\ell)$, $S_1^+(m_\ell), \theta = -\pi$ and $m = 1$,
D: Bounded by $m = 1$, $R_1^-(m_\ell)$ and $S_3^-(m_\ell)$,
E: Bounded by $m = 1, \theta = \pi$ and T.

When (m_r, θ_r) lies on $R_1^-(m_\ell)$, $R_3^+(m_\ell)$, $S_1^+(m_\ell)$ and $S_3^-(m_\ell)$, solution of the **Riemann problem** for WNLRT Eqs. (8.18) and (8.45) with initial data:

$$(m, \theta)(\xi, 0) = \begin{cases} (m_\ell, \theta_\ell = 0), & \xi \le 0 \\ (m_r, \theta_r), & \xi > 0, \end{cases} \tag{8.52}$$

are, respectively, a 1-R rarefaction wave, a 3-R rarefaction wave, a 1-S shock and a 3-R shock. Mapping these solutions to (x, y)-plane gives geometrical shapes of Ω_t is seen in Figs. 8.5 and 8.6.

Consider now a general Riemann problem with (m_r, θ_r), an arbitrary point in a relevant domain in Fig. 8.4. Let $P_r(m_r, \theta_r)$ be a point in the domain A. The solution of the Riemann problem exists because the curve $R_3^-(P_r)$ $(:= \theta - \sqrt{(8(m-1))} = \theta_r - \sqrt{(8(m_r - 1))},\ m < m_r)$, being below T, always meets $R_1^-(m_\ell)$ at a point at a point $P_i(m_i, \theta_i)$. In this case, the solution in upper half of (ξ, t)-plane consists of the constant state $(m_\ell, 0)$ on the left of a 1-R wave continuing up to an intermediate constant state (m_i, θ_i), which is the point of intersection P_i of the curves $R_1^-(m_\ell)$ and $R_3^-(P_r)$ and is unique. This intermediate constant state (m_i, θ_i) is on the left of a 3-R wave, which ends into the final constant state (m_r, θ_r). P_i has not been shown in the Fig. 8.4.

When the point $P_r(m_r, \theta_r)$ is a point in the domain A, the image of the above solution in (ξ, t)-plane to (x, y)-plane is shown in Fig. 8.7A, which shows that for $t > 0$, Ω_t consists of three straight parts joined by an elementary shape $\mathcal{R}_1$ moving down on Ω_t and an elementary shape $\mathcal{R}_3$ moving up on it. Symbolically, this is represented by

$$(m_r, \theta_r) \in A \to \mathcal{R}_1\mathcal{R}_3, \tag{8.53}$$

where the symbol $\mathcal{R}_1\mathcal{R}_3$ means that $\mathcal{R}_1$ is on lower part of Ω_t and $\mathcal{R}_3$ is on the upper part of Ω_t. $P_r(m_r, \theta_r) \in A$ can not give $\mathcal{R}_3\mathcal{R}_1$ since $\mathcal{R}_3$, moving upward, cannot be on lower part of Ω_t.

From the Fig. 8.4, it is simple to solve the Riemann problem also when P_r is in region B, C or D. For example, we can easily show that

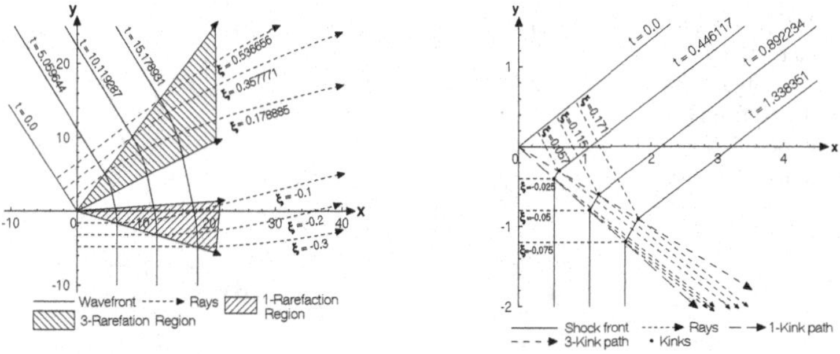

Fig. 8.7 8.7A: When P_r is in A, the shape of Ω_t is described by $\mathcal{R}_1\mathcal{R}_3$. 8.7B: When P_r is in C, the shape of Ω_t is described by $\mathcal{K}_1\mathcal{K}_3$. These figures have been reproduced from [7]- Figs. 10 and 12, with permission from the Publisher, Oxford University Press

$$(m_r, \theta_r) \in C \rightarrow \mathcal{K}_1\mathcal{K}_3. \tag{8.54}$$

In this case, the shape of Ω_t for $t > 0$ is shown in Fig. 8.7B. We see that Ω_t consists of three straight parts joined by a $\mathcal{K}_1$ moving down and $\mathcal{K}_3$ moving up on Ω_t. The rays enter the region bounded by the two kink lines from the two sides and become parallel.

We just mention the other two other results:

$$(m_r, \theta_r) \in B \rightarrow \mathcal{K}_1\mathcal{R}_3 \tag{8.55}$$

and

$$(m_r, \theta_r) \in D \rightarrow \mathcal{R}_1\mathcal{K}_3. \tag{8.56}$$

8.7 Interaction of Elementary Shapes

This is one of the most beautiful part of the applications of the KCL to nonlinear wavefront and shock front propagation. It has been discussed in great detail in [7] and not so completely in [42]. There are four elementary shapes[4] $\mathcal{R}_1$ elementary shape, $\mathcal{R}_3$ elementary shape, $\mathcal{K}_1$ kink and $\mathcal{K}_3$ kink, which propagate on a nonlinear wavefront Ω_t. Two elementary shapes, separated by a straight portion of Ω_t, may or may not interact. The process of interaction, if it takes place, may be instantaneous or may take finite or infinite time depending on the nature and relative strengths of the two elementary shapes. Although it is not possible to compute the shape of Ω_t during the process of interaction of non-zero duration without a full numerical solution of

[4]Neglecting the image of a contact discontinuity which have been removed by a proper choice of ξ.

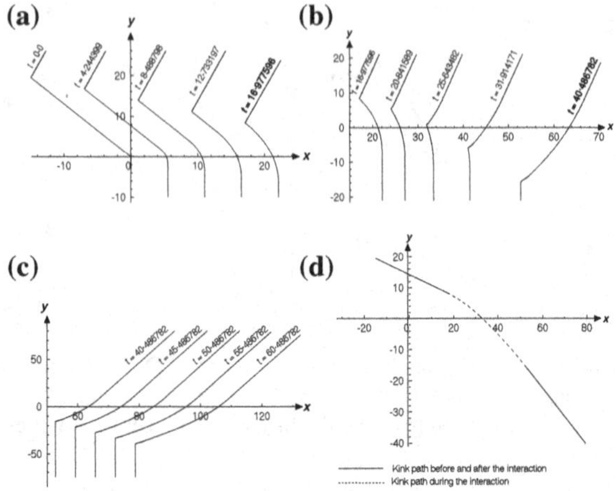

Fig. 8.8 $\mathcal{R}_1\mathcal{K}_1 \rightarrow \mathcal{K}_1\mathcal{R}_3$, when the interaction is complete. This figure has been reproduced from [7]-Fig. 15, with permission from the Publisher, Oxford University Press

the conservation laws (8.18) and relation (8.45), we can make a very good and exact prediction of the final results.

Consider initial value problem for (8.18) and (8.45) with Cauchy data

$$(m,\theta)(\xi,0) = \begin{cases} (m_\ell, \theta_\ell = 0), & -\infty < \xi \le \xi_\ell, \\ (m_i, \theta_i), & \xi_\ell < \xi \le \xi_r, \\ (m_r, \theta_r), & \xi_r < \xi \le \infty. \end{cases} \tag{8.57}$$

With the help of the above data and an appropriate choice of (m_i, θ_i) and (m_r, θ_r), we can produce any two elementary shapes on Ω_t separated by a straight part of Ω_t. Solving the above Cauchy problem in (ξ, t)-plane and mapping the solution in (x, y)-plane, we can study the interaction of any two elementary shapes.

We now use a notation $\mathcal{E}_\alpha\mathcal{E}_\beta$ to denote a shape of Ω_t consisting of an elementary shape $\mathcal{E}_\alpha$ joining states $(m_\ell, \theta_\ell = 0)$ and (m_i, θ_i) on the lower part of Ω_t and $\mathcal{E}_\beta$ joining (m_i, θ_i) and (m_r, θ_r) on the upper part of Ω_t. They are separated by a straight part of Ω_t with state (m_i, θ_i). Thus

$$\mathcal{R}_1\mathcal{K}_1 \rightarrow \mathcal{K}_1\mathcal{R}_3$$

means that interaction of $\mathcal{R}_1$ elementary shape and $\mathcal{K}_1$ kink on Ω_t will give rise to $\mathcal{K}_1$ kink and $\mathcal{R}_3$ shape on Ω_t.

First, we note that in $\mathcal{R}_1\mathcal{R}_1$ the two elementary shapes $\mathcal{R}_1$ and $\mathcal{R}_1$ moving down on Ω_t will not interact since the trailing end of $\mathcal{R}_1$ and the leading edge of $\mathcal{R}_1$ will move with the same velocity on ξ-axis. Similarly, the $\mathcal{R}_3$ and $\mathcal{R}_3$ moving up on Ω_t do not interact. We also note that in $\mathcal{K}_1\mathcal{K}_3$, $\mathcal{R}_1\mathcal{K}_3$, $\mathcal{R}_1\mathcal{R}_3$ and $\mathcal{K}_1\mathcal{R}_3$, the elementary shapes initially separated and moving in opposite directions on Ω_t do not interact.

Let us now take up the interaction $\mathcal{K}_3\mathcal{K}_1$. Though, this is the simplest it is also one of the most important interactions from the point of view of application, which we shall see very clearly in Fig. 9.4. For $\mathcal{K}_3\mathcal{K}_1$, given (m_ℓ, θ_ℓ), we have to choose the two states (m_i, θ_i) and (m_r, θ_r) and then find out the location of (m_r, θ_r) in Fig. 8.4. (m_i, θ_i) is the state on the right of a shock of the c_3 characteristic family with the state $(m_\ell, 0)$ on the left. Therefore, $(m_i, \theta_i) \in S_3^-(m_\ell)$. Similarly, we can see that $(m_r, \theta_r) \in S_1^+(m_i, \theta_i)$ so that $(m_r, \theta_r) \in C$ in Fig. 8.4. From (8.54) we get

$$\mathcal{K}_3\mathcal{K}_1 \rightarrow \mathcal{K}_1\mathcal{K}_3. \tag{8.58}$$

We could take up discussion of all remaining nine possible interactions: $\mathcal{K}_1\mathcal{K}_1$, $\mathcal{K}_3\mathcal{K}_3$, $\mathcal{R}_1\mathcal{K}_1$, $\mathcal{R}_3\mathcal{K}_3$, $\mathcal{K}_1\mathcal{R}_1$, $\mathcal{K}_3\mathcal{R}_3$, $\mathcal{R}_3\mathcal{R}_1$, $\mathcal{R}_3\mathcal{K}_1$ and $\mathcal{K}_3\mathcal{R}_1$. However some of these interactions require a lot of discussion, have many subcases and the interaction may not be complete in the sense that it may continue for infinite time. They have been discussed in great detail in [7] and we do not pursue them here. Before concluding this section, we present just one interesting result of interaction of $\mathcal{R}_1$ following $\mathcal{K}_1$ on Ω_t. This interaction may continue indefinitely, but when the interaction is complete we show the end result in Fig. 8.8:

$$\mathcal{R}_1\mathcal{K}_1 \rightarrow \mathcal{K}_1\mathcal{R}_3 \tag{8.59}$$

8.8 SRT: Eigenvalues of Equations and Interaction of Kinks

Equations (8.25) and (8.26) of SRT differ from the Eqs. (8.20) and (8.21) of WNLRT only in the addition source terms in (8.26). The eigenvalues of the SRT Eqs. (8.25)–(8.27) are

$$C_1 = -\sqrt{\frac{M-1}{2g^2}}, \quad C_2 = 0, \quad C_3 = 0, \quad C_4 = \sqrt{\frac{M-1}{2g^2}}. \tag{8.60}$$

By setting $t' = \frac{t}{\varepsilon}$, $0 < \varepsilon << 1$, we can show that on a small time scale t, i.e. $t' = O(1)$, the source terms in the two Eqs. (8.23) and (8.24) may be neglected and then (8.24) gets decoupled from other equations. Therefore, we find that the source terms do not play any role in the process of instantaneous interactions and the results of interaction of two kinks on a shock front are exactly same as those on a nonlinear wavefront. We shall see this important result in the numerical solution of SRT in Fig. 9.4 in the next chapter.

Chapter 9
2-D WNLRT and SRT—Some Applications

There are some technical details (see [8, 9, 42]) which we need while using the WNLRT and SRT to individual problems. Describing them would make the presentation quite long and take away our attention from the main results, which we wish to present very briefly. We shall discuss four types of applications below.

1. We first take up resolution of the linear caustic by genuine nonlinearity present in equations of a polytropic gas. This is a very important result discussed in Sect. 9.1 taking only the nonlinear waves produced by a wedged-shaped piston moving forward with a constant velocity. The geometry of a shock front produced by such a piston will be qualitatively the same.
2. Then, we discuss nonlinear waves and shock waves produced by an accelerating and decelerating piston. This problem has been discussed in a great detail in [8]. The calculations involved are quite intricate and we recommend that a reader goes through this section carefully. In Sect. 9.2, we present only the case of a convex wedged-shaped piston moving forward with a constant velocity. Unlike a concave piston problem, the solution in this case exists only under some conditions.
3. In Sects. 9.3 and 9.4, we discuss the evolution of a shock front of periodic shape and its corrugational stability. This is followed in Sect. 9.5 by a discussion of the evolution of shock front into a circle from a shape initially in form of a square-shaped closed curve. This is related to corrugational stability.
4. In the last section, we present the calculation of geometry of sonic boom produced by a simple manoeuvring aerofoil. This problem has been worked out by engineers with great interest due to its importance in the aviation industry. We show an approach which is mathematically elegant and computationally efficient with a very stable numerical scheme. The method is a great promise if it is extended to realistic situations—it is being pursued by S. Baskar.

P. Prasad, *Propagation of Multidimensional Nonlinear Waves and Kinematical Conservation Laws*, Infosys Science Foundation Series,
https://doi.org/10.1007/978-981-10-7581-0_9

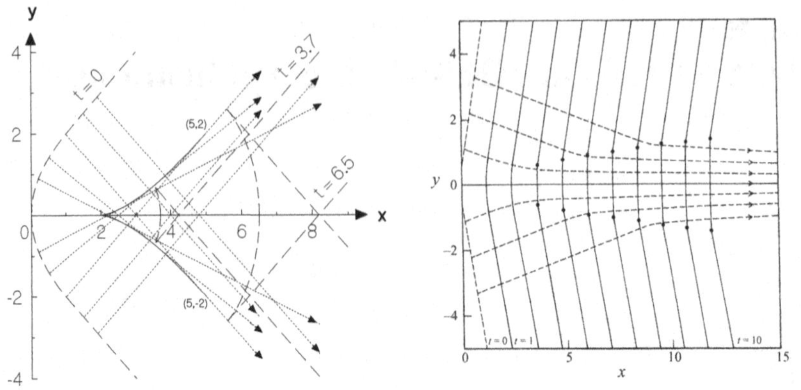

Fig. 9.1 Successive position of fronts starting from an initial shape wherein the central part is a parabola extended by their tangents on both sides. These figures have been reproduced from [32]—Fig. 1 and Fig. 2a, with permission from the Editor, Cambridge University Press.
Figure on left: Linear wavefront starting from Ω_0 given by (4.43) in an isotropic homogeneous medium with speed of propagation unity. - - - - : linear wavefront, —— : caustic, ... : rays.
Figure on right: Propagation of a weak shock front starting from Ω_0 with central part represented by $y^2 = 8x$ for $|y| < 1$. The initial amplitude distribution is given by $M_0 = 1.2$ and $\mathcal{V}_0 = 0.15$. - - - - : rays, —— : shock front Ω_t, kinks shown by dots on Ω_t. The kinks are formed between $t = 0$ and $t = 1$ and the result of interaction $\mathcal{K}_3\mathcal{K}_1 \rightarrow \mathcal{K}_1\mathcal{K}_3$ is not seen here as it happens quite early

9.1 Resolution of a Caustic

In 1976 Sturtevant and Kulkarni [54] reported some results of very carefully conducted experiments on the propagation of reflected 2-D shocks from a concave surface. The aim was to find geometrical shapes of concave shocks and shock strength distributions on it as shocks of increasing strength propagate. For a sound wave (i.e. a wave of infinitely small amplitude), it is well known that a converging wavefront folds in the caustic region with a cusp type of singularities. A very simple example is that of a converging wavefront with initial shape given by (4.43) and moving with a constant normal velocity $m = 1$ as seen in Fig. 9.1 on the left. It is a very interesting case where the caustic (shown by ——) is of finite extent from arête $(2, 0)$ to $(5, \pm 2)$). The wavefront folds after it reaches the arête at $t = 2$ and continues to fold beyond $(5, \pm 2)$ even after caustic has ended. Numerical solution (of balance equations (8.22)–(8.24) showing the successive position of a shock front starting from almost the same initial shape but with a constant shock strength $M = 1.2$ is shown in Fig. 9.1 on the right. A shape as in this figure has been captured experimentally in [54]. This corresponds to a moderately weak shock case[1] shown in Figs. 6 and 18 of [54] (with a slightly different initial front), where we see that the caustic is completely resolved.

[1]Called strong shock in [54].

There is also a beautiful figure from an exact solution of the WNLRT equations (8.18) and (8.45) in [42] (Fig. 6.3.2) showing resolution of the linear caustic due to genuine nonlinearity. Many very interesting results for successive positions of both converging and diverging shock fronts are given in Sect. 4 of [8] for a moderately strong shock. They all show complete resolution of the caustic.

Before we proceed further on the resolution of a caustic, we mention some results on the validity and accuracy of the SRT (which was derived also from the WNLRT, see Sect. 5.3.2):

1. The first attempt to verify and compare our method (2-D case) with other methods was done by Kevlahan [26] and later by us in [8]. Though Kevlahan used the NTSD in differential form, he followed a method by Whitham to locate the kinks (see [26], p. 180) and this method indeed gives the correct kinks location (see [42], p. 126). Kevlahan concludes '*The theory (based on equations* (8.25)–(8.27)) *is tested against known analytical solutions for cylindrical and plane shocks, and against a full direct numerical simulation (DNS) of a shock propagating into a sinusoidal shear flow. The test against DNS shows that the present theory accurately predicts the evolution of a moderately weak shock front, including the formation of shock-shocks due to shock focusing. The theory is then applied to the focusing of an initially parabolic shock and he finds that the shock shapes (by this theory) agreed well with the experimental results*'.
2. The second attempt to verify and compare our method (2-D case) with other methods is in [8], where the results of SRT have been compared with numerical solution of the full Euler equations and also with Whitham's shock dynamics. We especially refer to Figs. 1, 2, 8, 9 and 10 in [8]. The results show an excellent agreement of SRT results and Euler results.

Let us consider now a simple case of a nonlinear wavefront initially in the form of a wedge and with a constant distribution of wavefront velocity m

$$(m(\xi,0),\theta(\xi,0)) = \begin{cases} (m_\ell,\theta_\ell) & \xi<0 \\ (m_\ell,-\theta_\ell). & \xi>0 \end{cases} \tag{9.1}$$

For $0<\theta_\ell<\frac{\pi}{2}$, the initial wavefront is concave as shown in Fig. 9.2 and for $-\frac{\pi}{2}<\theta_\ell<0$ it is convex.

The method of construction of the linear wavefront with the help of rays starting from an initial Ω_0 is explained in Fig. 9.2. We are now concerned with a nonlinear wavefront. The solution of the Riemann problem for Eqs. (8.18) and (8.45) with initial data (9.1) for $0<\theta_\ell<\frac{\pi}{2}$ is quite simple. The point $(m_\ell,-\theta_\ell)$ falls in the region C of Fig. 8.4 (the figure has to be shifted up by taking P_ℓ on the line $\theta=\theta_\ell$). Due to symmetry, the solution in (ξ,t)-plane consists of

$$(m,\theta)(\xi,t) = \begin{cases} (m_\ell,\theta_\ell), & -\infty<\xi\le -St, \\ (m_i,0), & -St<\xi\le St, \\ (m_\ell,-\theta_\ell), & St<\xi\le\infty, \end{cases} \tag{9.2}$$

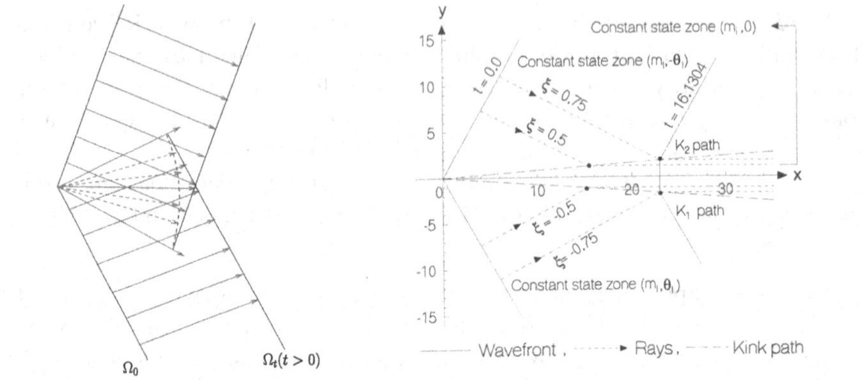

Fig. 9.2 Figure on the left **a**: Linear wavefront which starts with a concave wedged shape. ..... wavefront produced by the corner O by Huygen's method. This is an arc of a circle with centre at O. —— wavefront produced by the smooth part of the initial wavefront by ray theory or Huygen's method. Figure on the right **b**: Nonlinear wavefront produced by an initially concave wedged-shaped Ω_0 reproduced from [42]

where S is the shock velocity of a shock of the third characteristic family. We need to find out only m_i in the intermediate domain $-St < \xi \leq St$.

We use (8.38) for a shock in c_1 characteristic family with $\theta_r = 0$ and $m_r = m_i$. Similarly, we use it for a shock in c_3 characteristic family with $\theta_\ell = 0, \theta_r = -\theta_\ell$ and $m_l = m_i$, $m_r = m_\ell$. Both these operations give the same result

$$m_i g(m_i) + m_l g(m_l) = (m_i g(m_l) + m_l g(m_i)) \cos\theta_l, \tag{9.3}$$

which is the equation we need to solve for m_i for a given m_ℓ and θ_ℓ in (9.1). When we map the solution (9.2) to (x, y)-plane as explained in Sect. 8.3, we get the Fig. 9.2B. This solution shows that the linear caustic in Fig. 9.2A is completely resolved by the genuine nonlinearity of the Euler equations. The WNLRT and hence the Eq. (9.3) is valid for $0 < m - 1 << 1$. Recently, we[2] have carefully drawn the Fig. 8.4 for very small positive values of $m_\ell - 1$ and many values of θ_ℓ, $0 < \theta_\ell < \frac{\pi}{2}$ and found that the state $(m_\ell, -\theta_\ell)$ always lies in the region C. We have solved (9.3) numerically for m_i and got a value of m_i for all these values of m_ℓ and θ_ℓ. Thus, we find that even for very small positive values of $m_\ell - 1$, the linear caustic is resolved and nonlinear wavefront has no fold but has two kinks as seen in the Fig. 9.2B.

From the discussion in the paper [54], we may think that for a given initial geometry of the front, there exists a critical value m_c of m such that for $m < m_c$ the front folds with a geometry similar to a linear front (as in Fig. 9.1A) and for $m > m_c$ the caustic is resolved and the front develops two kinks. The result discussed above shows, at least for a wedged-shaped initial shape (9.1), no such critical m_c exists and caustic does not appear. In other cases also when the geometry of the initial front is

[2]Baskar and Prasad - 2016, unpublished.

as in the Fig. 9.1A, if there exists such a m_c, then for $m < m_c$ when the rays will first meet at the arête and cross, the curvature of the front will tend to infinity in the first approximation and finite but too large in the second approximation [11, 30]. This means that the amplitude of the wave, will tend to infinity according to Eq. (5.53) or (5.56), even for a very week nonlinear wave or shock. Thus at the arête full nonlinear effect will come into play. We believe that in the Figs. 6b and 18b of [54], two kinks appear as in Fig. 9.2B. We further believe that shock–shock paths seen in Figs. 6(b,c) and 18(b,c) of [54] show the interaction of two shocks according to the result (8.58) and seen in the Fig. 5 of [32] and reproduced in Fig. 9.4.

We conclude in this section that even very small amplitudes of a nonlinear wavefront and a shock front resolve the caustic and cusp type of singularities do not appear on converging fronts.

9.2 Nonlinear Wavefront Produced by a Convex Piston

Consider a wedged-shaped piston which is initially at rest and then suddenly starts moving into the gas ahead with a constant velocity $u_0 > 0$ in the direction of the symmetry, assumed to be the direction of the x-axis. Let ξ be the distance along the piston measured from the vertex. Then for $t > 0$ and $u_0 > 0$, the piston position is given by

$$x_p(\xi, t) = \begin{cases} -\xi \sin \Theta_0 + u_0 t, & \xi > 0 \\ \xi \sin \Theta_0 + u_0 t, & \xi < 0 \end{cases}, \qquad y_p(\xi, t) = \xi \cos \Theta_0. \tag{9.4}$$

We note that[3] for a convex piston $0 < \Theta_0 < \frac{\pi}{2}$.

The nonlinear wavefront (also shock front) produced by the motion of a piston is coincident with the piston at $t = 0$. Even for a more general piston motion than that in (9.4), i.e. accelerating piston we can deduce the expressions for initial values m_0 (for a shock front M_0 and $\mathcal{V}_0$), see Sects. 3 and 5 in [8]. For this piston motion, we get the following initial value for WNLRT,

$$m_0(\xi) = 1 + \frac{1}{2}(\gamma + 1)u_0 \cos \Theta_0. \tag{9.5}$$

For a given u_0 and Θ_0, this value $m_0(\xi)$ is constant.

In the previous section, for a concave wedged- shaped piston moving into the gas ahead, (see also Sect. 5.1 in [8]), we always got a nonlinear wavefront Ω_t whatever may be the value of u_0 and Θ_0. However, for a convex piston there is exist no solution of the initial value for the WNLRT equations with initial data (9.5) unless the initial piston velocity u_0 is sufficiently large satisfying (for derivation see Sect. 5.2.1 in [8])

[3]Here Θ_0 is negative of Θ_0 in Sect. 5.2.1 of [8].

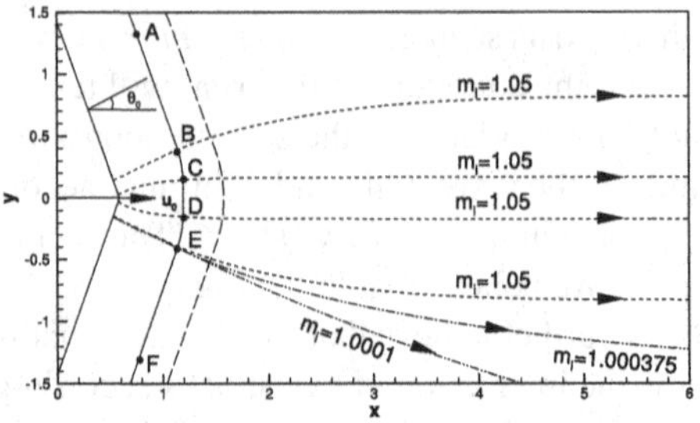

Fig. 9.3 The nonlinear wavefront (shown by solid line) ABCDEF at $t = 0.6134$ for $m_0 = 1.13$ and $\Theta_0 = \pi/8$ (which gives $m_i = 1.05$) consists of two curved parts BC and DE separating a straight central disc CD from the two straight infinite wings BA and EF. For the same Θ_0, if m_0 is chosen so that $m_i = 1.0001$, then the point C and D almost coincide (wavefront shown by long broken lines at $t = 0.9927$) and the rays (shown by $- \cdot\cdot - \cdot\cdot$) becomes almost straight lines as in the case of linear rays. This figure has been reproduced from [8]—Fig. 5, with permission from the Editor, Cambridge University Press

$$u_0 > \frac{{\Theta_0}^2}{4(\gamma+1)\cos\Theta_0}, \tag{9.6}$$

where we note that the right-hand side is positive.

To find the successive positions of the nonlinear wavefront when condition (9.6) is satisfied, we need to solve a Riemann problem for the WNLRT equations (8.18), (8.45) with initial data

$$(m(\xi,0),\theta(\xi,0)) = \begin{cases} (m_0,-\Theta_0) & \xi<0 \\ (m_0,\Theta_0) & \xi>0. \end{cases} \tag{9.7}$$

The solution of the above Riemann problem at any time t consists of three constant states $(m_0, -\Theta_0)$ for $-\infty < \xi < c_1(m_0)t$; $(m_i, 0)$ for $c_1(m_i)t < \xi < c_2(m_i)t$ and (m_0, Θ_0) for $c_2(m_0)t < \xi < \infty$. The constant states are joined by sections of two simple waves at t. The value of m_i can be easily obtained and is given in [8] (Eq. (5.10)). Figure 9.3 shows a nonlinear wavefront corresponding to $m_0 = 1.13$ and $\Theta = \pi/8$. It also contains a number of rays corresponding to different values of m_0 and caption describes some of the interesting features from [8].

9.3 Evolution of a Shock Front of Periodic Shape

We present here a few results of successive positions of shock fronts whose shapes are initially periodic, and of course their shapes remain periodic as they evolve. The fulldetails of the numerical methods of the solution and details of many results are given in [32]. We briefly mention in the caption of the figure the main points and give the beautiful geometrical shapes as the shocks evolve. We show formation and propagation of kinks on these shocks and interaction of kinks according to the result (8.58).

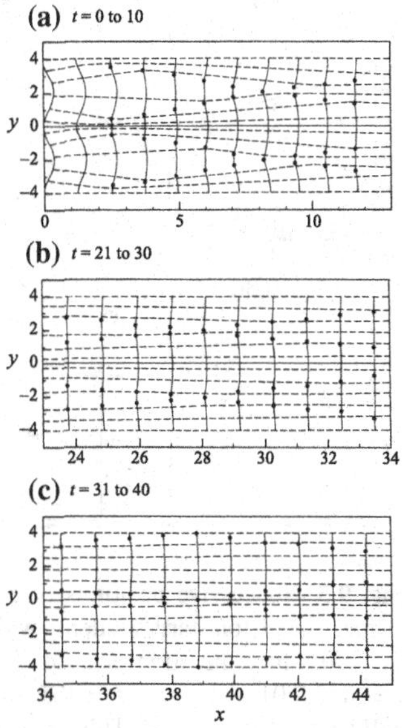

Fig. 9.4 Successive positions of an initially sinusoidal shock front (shown by continuous line) plotted at $t = 0, 1, 2, 3, \cdots$, 40 and rays (shown by broken lines). In the figure, we see $\mathcal{K}_3$ moving upwards and $\mathcal{K}_1$ moving downwards resulting in interaction results $\mathcal{K}_3\mathcal{K}_1 \rightarrow \mathcal{K}_1\mathcal{K}_3$ happening many times. The shock has become almost straight and rays parallel to x-axis from $t = 31$ to 40. This figure have been reproduced from [32]—Fig. 5, with permission from the Editor, Cambridge University Press

Remark 9.3.1 The differential form of SRT equations in [32] is the same as Eqs. (8.25)–(8.27). However, the balance equations representing the two closure relations of KCL, namely (3.10) and (3.11) in [32] are different from (8.23) and (8.24) but those in [8] are same as (8.23) and (8.24). Some of the results in [32] are not only qualitatively the same but also quite close to those obtained from the SRT equations (8.22) and (8.24).

We first take the initial shock front Ω_0 to be in a periodic sinusoidal shape

$$x = 0.2 - 0.2\cos\left(\frac{\pi y}{2}\right). \tag{9.8}$$

We choose the shock Mach number $M_0 = 1.2$ and $\mathcal{V} = 0.1$ on Ω_0. The shape of the shock at various times are given in Fig. 9.4.

9.4 Corrugational Stability

The corrugational stability of a front is defined as the stability of a planar front to perturbations. This means that the perturbations in the shape of a planar front ultimately disappear as time tends to infinity and the front will tend to be a plane. This topic has been discussed in [8, 32] in 2-D and in [2, 5] in 3-D. Let us just reproduce some figures from [8, 32]. We shall discuss the 3-D case in more detail in the next chapter.

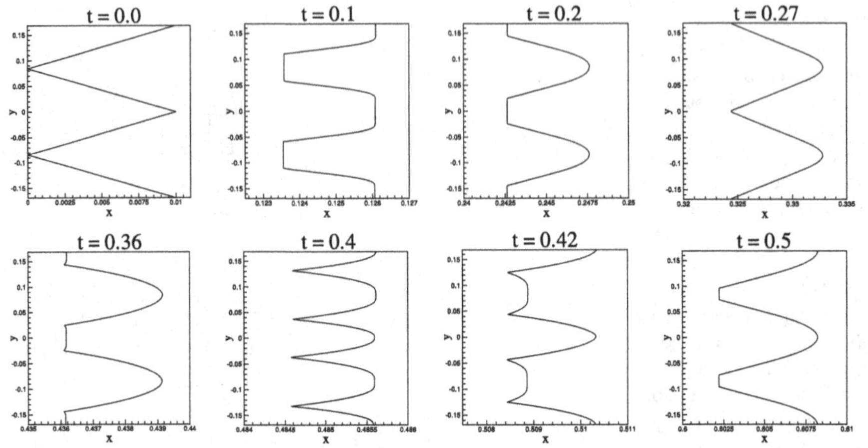

Fig. 9.5 Successive positions of a shock front produced by an accelerating piston of periodic shape. The piston starts suddenly with a velocity $u_0 = 0.333$ and then accelerates with acceleration $u_1 = 0.5$ in x-direction (see Eq. (5.1) in [8]). The amplitude of oscillation of the shock front at $t = 0$ is 0.01 and that at $t = 0.5$ it is approximately equal to 0.005. The amplitude of oscillation will tend to 0 as $t \to \infty$. This figure has been reproduced from [8]—Fig. 15, with permission from the Editor, Cambridge University Press

One of the finest examples of corrugational stability is seen in Fig. 9.4. In this, the amplitude of depression and elevation in the shape of the shock front keeps on decreasing and finally the front tends to become plane.

In [8], we have the case of a shock produced by an accelerating piston of periodic shape in y-direction. The piston suddenly starts moving in x-direction with velocity u_0 and accelerates with a constant acceleration u_1 in the same direction. Between $y = -0.15$ and $y = 0.15$, there are two concave wedges and and a convex wedge, and the piston moves suddenly with a nonzero velocity and thereafter with a constant acceleration in x-direction. The shock front, produced by the piston, will initially coincide with the piston and initial values of both quantities, namely shock Mach number $M_0(\xi)$ and the gradient $\mathcal{V}_0(\xi)$ can be calculated from the velocity u_0 and acceleration u_1 of the piston. Details are available in Sects. 3 and 5 in [8]. Here, we reproduce two figures, the first Fig. 9.5 shows corrugational stability and the second Fig. 9.6 gives comparison of the shock positions by our SRT and Whitham's shock dynamics. Figure 9.6 also shows that the nonlinear wavefront moves far ahead of the shock front.

9.5 Evolution of a Shock from Closed Curves into Circles

This is one of two very interesting applications of our WNLRT and SRT, discussed in great detail in Sect. 6 of [8]. We take a blast wave produced by an explosive placed in a closed 2-D container of any shape—for simplicity let the container be in the

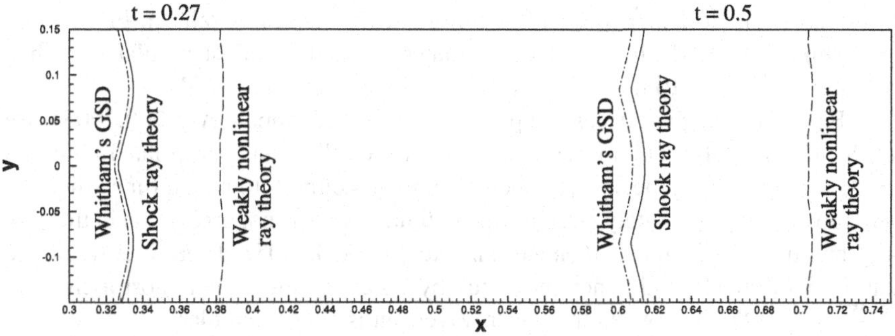

Fig. 9.6 Comparison of results in the case of a periodic shock front produced by an accelerating piston with initial velocity $u_0 = 0.333$ and acceleration $u_1 = 0.5$. This figure has been reproduced from [8]—Fig. 16, with permission from the Editor, Cambridge University Press

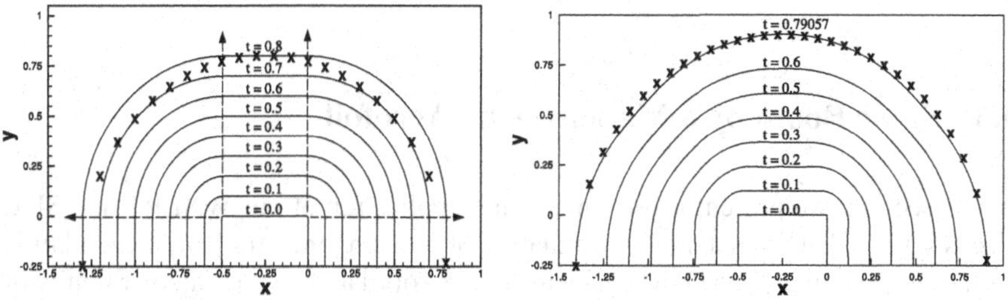

Fig. 9.7 **a** Successive positions at times t of a leading wavefront from a blast wave due to a square-shaped source using linear theory. A circle (shown by the symbol 'X') of an appropriate radius and centre at the centre of the square has been compared with the linear wavefront at t = 0.8. **b** Successive positions of a leading shock front (at different times t) from a blast wave due to a square-shaped source using shock ray theory with $u_0 = 0.333$, $u_1 = 0.5$. A circle of an appropriate radius and centre at the centre of the square has been shown by the symbol 'X' to compare with the shock front at $t = t_c$. Here $M_0 = 1.2$ and $\mathcal{V}_0 = -0.28125$. These figures have been reproduced from [8]—Figs. 11 and 12, with permission from the Editor, Cambridge University Press

shape of a square. *Our main aim is to show that the shock wave, which is produced initially in the shape of the container tends to acquire a circular shape very rapidly due to genuine nonlinearity of the Euler equations.* In this process, the reference [8] shows an additional result that the SRT gives position and shape of the shock very close to that obtained by numerical solution of the Euler equations, which we skip. A linear wavefront produced by a square-shaped piston and moving with the velocity $m = 1$ will take infinite time to acquire a circular shape. As seen in Fig. 9.7a, at any time it will have four straight parts (obtained by joining the tips of straight rays in normal directions from the sides of the squares) joined by circular arcs with centres at the corners of the square (obtained by Huygens' method). In [8], the following argument is given to approximately estimate a critical t_{lc}, when the linear wavefront becomes a circle: 'If a is the length of the side of the square at $t = 0$, the ratio of total length $4a$ of the straight parts of the linear wavefront to that of the circular arcs

$2\pi t$ is $2a/(\pi t)$. Here $a = 0.5$ and, therefore, at $t = 0.8$, this ratio in Fig. 9.7A is approximately $1/3$. The linear wavefront may be treated as almost circular, say when this ratio is $\frac{2a}{\pi t_{lc}} = 1/10$ i.e. $t_{lc} = O\,(20a/\pi)$ which is equal to ≈ 3'.

In Fig. 9.7B, we give successive positions of the SRT shock. At $t = 0.79057$, we see that the shock front almost coincides with the circle of an appropriate radius and centre at the origin of the square. Therefore, if we define t_c to be the time when the SRT shock is approximately circular (the definition of t_c is not precise, as in the case of t_{lc} but just an indication), then we can take t_c to be 0.79057. Therefore, we notice that the SRT shock (and hence the shock by Euler equations) is almost a circle in $\frac{0.79057}{3} \approx \frac{1}{3}$ of the time when the linear wavefront is almost circular.

On the shock front, there are nonlinear waves which move with the characteristic velocities (8.59). These waves cause the depressions and elevations to spread over the shock front leading to corrugational stability and consequently a closed shock front tends to become a circle very rapidly.

9.6 Sonic Boom by a Maneuvering Aerofoil

In this section, we present a very interesting application of our WNLRT and SRT, discussed in [9]. This is a novel formulation of reducing the problem of calculation of the sonic boom produced by a maneuvering aerofoil as a one- parameter family of Cauchy problems. The system of equations governing the geometry of shocks and also that of nonlinear waves is hyperbolic, when they originate from the front part of the aerofoil and has elliptic nature elsewhere, showing that unlike the leading shock, the trailing shock is always smooth.

9.6.1 Description and Basic Assumptions

Consider a 2-D unsteady flow produced by a thin maneuvering supersonic aerofoil either accelerating on a straight path or moving on a curved path. We are interested in calculating the sonic boom produced by the aerofoil, the point of observation being far away say at a distance L, from the aerofoil. We use coordinates x, y and time t non-dimensionalized by L and the sound velocity a_0 in the ambient medium. In a local rectangular coordinate system (x', y') with origin O' at the nose of the aerofoil and $O'x'$ axis tangential to the path of the nose, which moves along a curve $(X_0(t), Y_0(t))$, let the upper and lower surfaces of the aerofoil be given by

$$(x' = \zeta, y' = b_u(\zeta)) \text{ and } (x' = \zeta, y' = b_l(\zeta)), -d < \zeta < 0 \qquad (9.9)$$

respectively. Here d is the non-dimensional camber length. We assume that $b_u'(-d) > 0$, $b_u'(0) < 0$, $b_l'(-d) < 0$ and $b_l'(0) > 0$, so that the nose and the tail of the aerofoil are not blunt. We further assume that (i) the aerofoil is thin and (ii) the cumber length

$\bar{d}$ is small compared with L and the radius of curvature of the path of the aerofoil and (iii) cumber line is aligned with the direction of motion of the aerofoil. With $d = \frac{\bar{d}}{L}$, these assumptions imply

$$d = \frac{\bar{d}}{L} = O(\epsilon), \quad O\left\{\frac{max_{-d<\zeta<0}b_u(\zeta)}{d}\right\} = O\left\{\frac{max_{-d<\zeta<0}(-b_l(\zeta))}{d}\right\} = O(\epsilon), \tag{9.10}$$

where ϵ is a small positive number. Then, the amplitude w of the perturbation in the sonic boom also satisfies $w = O(\epsilon)$. w is related to $\tilde{w}$ in (5.10) by $w = \varepsilon\tilde{w}$.

We show the geometry of the aerofoil and the sonic boom produced by it at a time t in Fig. 9.8. The sonic boom produced either by the upper or lower surface consists of a leading shock LS: $\Omega_t^{(0)}$ and a trailing shock TS: $\Omega_t^{(-d)}$ and since high-frequency approximation is satisfied by the flow between the two shocks, a one-parameter family of nonlinear wavefronts $\Omega_t^{(\zeta)}(-d < \zeta < 0, \zeta \neq G)$ originating from the points P_ζ on the aerofoil in between the two shocks. *Note that the leading shock* $\Omega_t^{(0)}$, *the nonlinear wavefront* $\Omega_t^{(\zeta)}$ *and trailing shock* $\Omega_t^{(-d)}$ *all originate from fixed points of the aerofoil and move with it.* The nonlinear wavefronts produced from the points on the front portion of the aerofoil start interacting with the LS $\Omega_t^{(0)}$ and those from the points near the trailing edge do so with the TS $\Omega_t^{(-d)}$, and after the interaction they keep on disappearing continuously from the flow. These two sets, one interacting with LS and another interacting with TS are separated by a linear wavefront $\Omega_t^{(G)}$, which originates from a point P_G where the function $b_u(\zeta)$ ($b_l(\zeta)$) is maximum (minimum). Figure 9.8 (part 2) shows an enlarged version of the upper part of the Fig. 9.8 (part 1) near the aerofoil.

9.6.2 Introduction of a Ray Coordinate System

We first introduce a ray coordinate system (ξ, t) for the nonlinear wavefront $\Omega_t^{(\zeta)}$. The front $\Omega_t^{(\zeta)}$ at a given time t can be obtained joining of the tips of the rays (at time t) in (x, y)-plane starting from positions $P_\zeta|_\eta$ of P_ζ at times $\eta < t$ as shown in part 3 of the Fig. 9.8. Therefore, a value of η identifies a ray and we choose

$$\xi = -\eta, \quad \eta \leq t \tag{9.11}$$

for $\Omega_t^{(\zeta)}$ from the upper surface (for lower surface we need to choose $\xi = \eta, \eta \leq t$). When $\xi \equiv -\eta = -t$, the points A, B and C in the part 3 of the Fig. 9.8 coincide. Hence, the base point P_ζ of $\Omega_t^{(\zeta)}$, which lies on the upper surface of the aerofoil, corresponds to a point, which lies on the line $\xi + t = 0$ in the (ξ, t)-plane. Where does a point C on $\Omega_t^{(\zeta)}$ lie in (ξ, t)-plane? The ray, which reaches C starts from a point $P_{\zeta|\eta}$, where $\eta < t$, i.e., $0 < -\eta + t$. Hence at C, $\xi + t = -\eta + t > 0$.

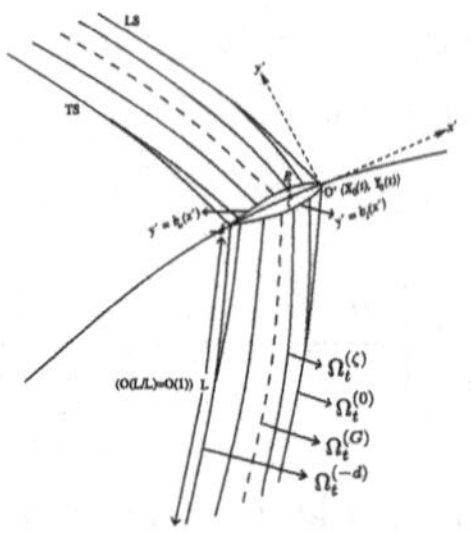

Figure 1: *Sonic boom produced by the upper and lower surfaces: $y' = b_u(x')$ and $y' = b_l(x')$ respectively. The boom produced by either surface consists of a one parameter family of nonlinear wavefronts*

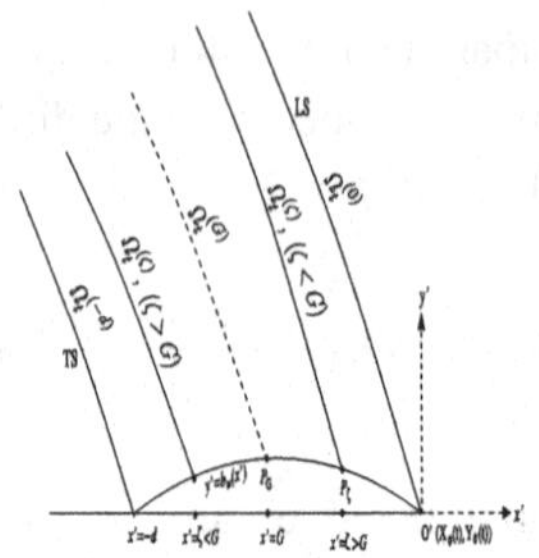

Figure 2: *It An enlarged version of the upper part of the Figure 1 near the aerofoil.*

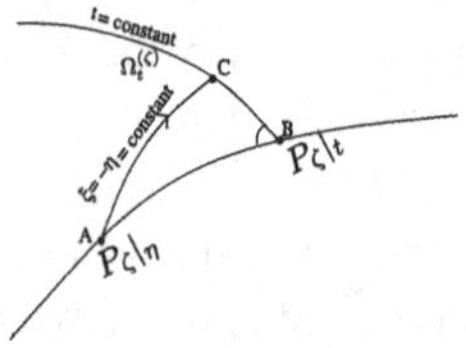

Figure 3: *A formulation of the ray coordinate system (ξ, t) for $\Omega_t(\zeta)$. AB represents the path of a fixed point P_ζ on the aerofoil. A is the position of P_ζ at time η and B that at time t.*

Fig. 9.8 Figs. 1, 2 and 3 describe the formulation of the sonic boom problem. This figure has been reproduced from [9]— Fig. 2.1, 2.2 and 2.3 with permission from the Indian Academy of Sciences

The ray coordinates for the shock $\Omega_t^{(0)}$ or $\Omega_t^{(-d)}$ can be formulated exactly the same way.

Equations governing $\Omega_t^{(\zeta)}$: For $-d < \zeta < 0$, each of the curves $\Omega_t^{(\zeta)}$ are nonlinear wavefronts with ray coordinates (ξ, t) and they satisfy KCL (8.18) with (8.19). Similarly, the leading shock $\Omega_t^{(0)}$ and trailing shock $\Omega_t^{(-d)}$ satisfy the SRT equations (8.22)–(8.24). To find the solution in $\xi + t > 0$, we need Cauchy data on the curve $\xi + t = 0$.

9.6.3 Cauchy Data in Ray Coordinates

These are derived from the boundary conditions of the inviscid flow condition on the surface of the aerofoil, which imposes only one condition that the fluid velocity in normal direction at a point on the surface is equal to the normal component of the velocity of the surface at that point.

Cauchy data of m, g and θ for a nonlinear wavefront $\Omega_t^{(\zeta)}$ from the upper surface: The nonlinear wavefront $\Omega_t^{(\zeta)}$, $(-d < \zeta < 1, \zeta \neq G)$ is governed by the system (8.18) and (8.19). The Cauchy data for this nonlinear wavefront comes from the inviscid flow condition at the point $P_{\zeta|t}$, which lies on $\xi + t = 0$ in the (ξ, t)-plane. This involved some very difficult procedure given in [9], which we omit here. Retaining only the leading order terms, we finally get

$$m(\xi, -\xi) = m_0(\xi) := 1 - \frac{(\gamma+1)(\dot{X}_0^2 + \dot{Y}_0^2)b_u'(\zeta)}{2(\dot{X}_0^2 + \dot{Y}_0^2 - 1)^{1/2}}, \tag{9.12}$$

$$g(\xi, -\xi) = g_0(\xi) := (\dot{X}_0^2 + \dot{Y}_0^2 - 1)^{1/2}, \tag{9.13}$$

and

$$\theta(\xi, -\xi) = \theta_0(\xi) := \frac{\pi}{2} + \psi - \sin^{-1}\{1/(\dot{X}_0^2 + \dot{Y}_0^2)^{1/2}\}, \tag{9.14}$$

where $\psi = \tan^{-1}\left\{\dot{Y}_0/\dot{X}_0\right\}$. Since $b_u'(\zeta) < 0$ for $G < \zeta < 1$ and $b_u'(\zeta) > 0$ for $-d < \zeta < G$, $m_0 > 1$ on P_ζ for $G < \zeta < 1$ and $m_0 < 1$ on P_ζ for $-d < \zeta < G$. This can be used to argue that

$$m > 1 \text{ on } \Omega_t^{(\zeta)}, G < \zeta < 0 \text{ and } m < 1 \text{ on } \Omega_t^{(\zeta)}, -d < \zeta < G. \tag{9.15}$$

Since the eigenvalues of the system (8.18) and (8.19) governing evolution of $\Omega_t^{(\zeta)}$ are given by (8.34), we get a Cauchy problem for a hyperbolic system for $\Omega_t^{(\zeta)}$, $G < \zeta < 0$ and an elliptic system for $\Omega_t^{(\zeta)}$, $-d < \zeta < G$ (we call it elliptic even though $\lambda_2 = 0$ is real).

Cauchy data of M, G, Θ and $\mathcal{V}$ for the trailing and leading shock fronts $\Omega_t^{(-d)}$ and $\Omega_t^{(0)}$ from the upper surface: The derivation of the Cauchy data on $\xi + t = 0$ for the system (8.22)–(8.24) governing the evolution of the leading and trailing shock fronts is even more complex. We quote from [9] the leading order terms in this Cauchy data

$$M(\xi, -\xi) = M_0(\xi) := 1 - \frac{(\gamma+1)(\dot{X}_0^2 + \dot{Y}_0^2)b_u'(\xi)}{4(\dot{X}_0^2 + \dot{Y}_0^2 - 1)^{1/2}} \tag{9.16}$$

$$G(\xi, -\xi) = G_0(\xi) := (\dot{X}_0^2 + \dot{Y}_0^2 - 1)^{1/2} \tag{9.17}$$

$$\Theta(\xi, -\xi) = \Theta_0(\xi) := \frac{\pi}{2} + \psi - \sin^{-1}\{1/(\dot{X}_0^2 + \dot{Y}_0^2)^{1/2}\} \tag{9.18}$$

$$\mathcal{V}(\xi, -\xi) = \mathcal{V}_0(\xi) := \frac{\gamma+1}{4}\{\Omega_{P(-d)} w_0(\xi) - \mathcal{F}(-d, t)\} \tag{9.19}$$

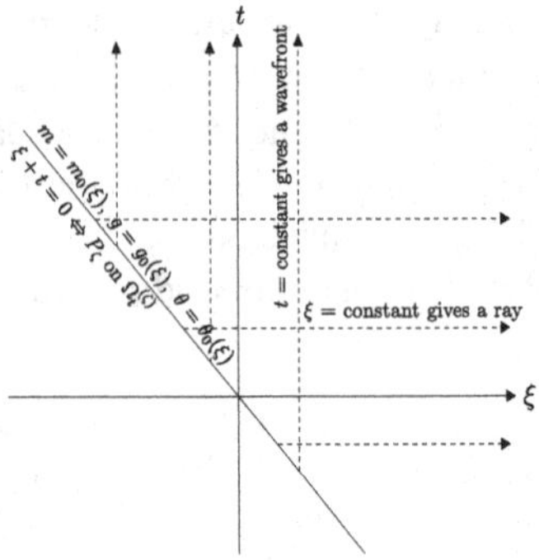

Fig. 9.9 The successive positions of a fixed point P_ζ, $\zeta \neq G$ on the aerofoil corresponds to points on $\xi + t = 0$, where Cauchy data is prescribed. Solution domain is $\xi + t > 0$. This figure has been reproduced from [9]—Fig. 4.3 with permission from the Indian Academy of Sciences

where

$$\begin{aligned}
\Omega_{P_{(-d)}} &= \frac{(\dot{X}_0\ddot{X}_0 + \dot{Y}_0\ddot{Y}_0)}{2G_0(\dot{X}_0^2 + \dot{Y}_0^2)(\dot{X}_0^2 + \dot{Y}_0^2 - 1)^{1/2}} + \frac{\dot{X}_0\ddot{Y}_0 - \dot{Y}_0\ddot{X}_0}{G_0\dot{X}_0^2} \\
\mathcal{F}(\zeta, t) &= \frac{(\dot{X}_0^2 + \dot{Y}_0^2)b_u''(\zeta)}{(\dot{X}_0^2 + \dot{Y}_0^2 - 1)^{1/2}}\{\dot{\mathcal{X}}_0(t)\} - \frac{(\dot{X}_0^2 + \dot{Y}_0^2 - 2)(\dot{X}_0\ddot{X}_0 + \dot{Y}\ddot{Y}_0)}{(\dot{X}_0^2 + \dot{Y}_0^2 - 1)^{3/2}}b_u'(\zeta) \\
\mathcal{X}_0 &= X_0\cos\psi + Y_0\sin\psi
\end{aligned}$$

On the leading shock $\Omega_t^{(0)}$, $M > 1$, hence the eigenvalues (8.59) of the SRT equations are real and for the determination of $\Omega_t^{(0)}$ we get a the Cauchy problem for a hyperbolic system (8.22)–(8.24). On the trailing shock $\Omega_t^{(-d)}$, $M < 1$, the system has two eigenvalues C_1 and C_4, which are purely imaginary, we call it elliptic (though not strictly).

Solution and construction of $\Omega_t^{(\zeta)}$, $-d \leq \zeta \leq 0$: We have reduced the problem of finding the state of the fluid on the leading shock $\Omega_t^{(0)}$, the trailing shock $\Omega_t^{(-d)}$ and the nonlinear wavefronts in between these two to a one-parameter family of Cauchy problems. In each problem we determine the geometry and the Mach number the front $\Omega_t^{(\zeta)}$, $-d \leq \zeta \leq 0$, which is either a nonlinear wavefront or a shock front. For each $\Omega_t^{(\zeta)}$ the data is prescribed on the line $\xi + t = 0$ and the solution domain is the half plane $\xi + t > 0$ as shown in Fig. 9.9.

Once a cauchy problem is solved (numerically), we use the first two of the Eq. (8.15), i.e. $x_t = m\cos\theta$, $y_t = m\sin\theta$ or $x_t = M\cos\theta$, $y_t = M\sin\theta$ to map the solution from (ξ, t)-plane to (x, y)-plane. A line parallel to t-axis in Figure (9.9) is mapped onto a ray and a line parallel to ξ-axis is mapped onto the $\Omega_t^{(\zeta)}$. This procedure gives the complete geometry of a front $\Omega_t^{(\zeta)}$ and its Mach number m (for a nonlinear wavefront $= 1 + \varepsilon\frac{\gamma+1}{2a_0}\tilde{w}$) or M (for shock front $= 1 + \varepsilon\frac{\gamma+1}{4a_0}\tilde{w}$), see (5.50). Once the amplitude $\tilde{w}$ is calculated in (x, y)-plane, the state of the fluid is determined with the help of Eq. (5.10) at the points of $\Omega_t^{(\zeta)}$.

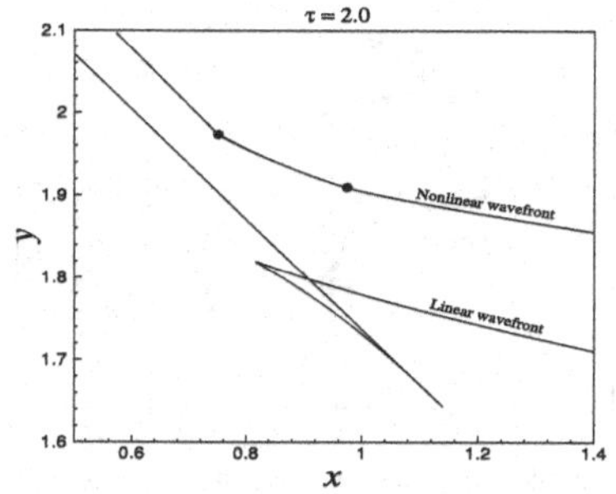

Figure 4: *Sonic boom wavefront at t = 2 from the leading edge of an accelerating aerofoil moving in a straight path. Kinks on the nonlinear wavefront are shown by dots. The initial Mach number is 1.8 and the acceleration is 10 in the line interval (1, 1/2).*

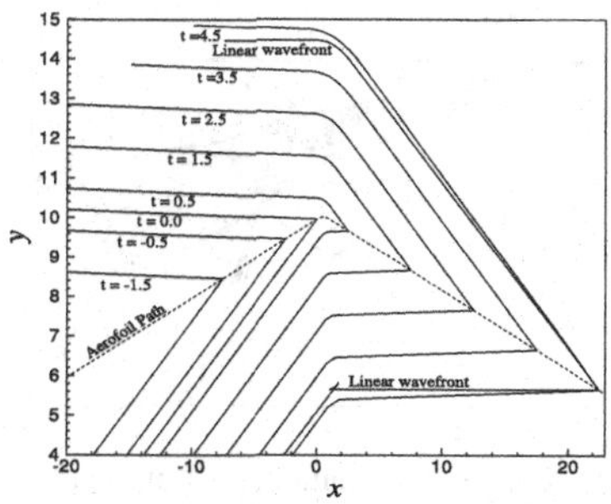

Figure 5: *The nonlinear wavefront from the leading edge of an aerofoil moving with a constant Mach number 5 along a path concave downwards with* $b'_u(0) = -0.01$.

Fig. 9.10 Parts 4 and 5 describe two results of the sonic boom. This figure has been reproduced from [9]—Figs. 5.1 and 5.2 with permission from the Indian Academy of Sciences

The most interesting result seen from our new formulation of the sonic boom problem is the elliptic nature of the equations governing the trailing shock $\Omega_t^{(-d)}$. This implies that whatever may be the flight path and acceleration of the aerofoil, the trailing shock $\Omega_t^{(-d)}$ must be smooth. All these features, which we obtain from our theory are seen in the Euler's numerical solution of [24]. Just two of the results have been reproduced in Fig. 9.11 with the permission of the authors.

9.6.4 Results

We present some results in Parts 4 and 5 of the Fig. 9.10. From numerical computation, we find that the geometric shape of the nonlinear wavefront is not only topologically same as that of the LS $\Omega_t^{(0)}$ but is very close to it. Hence, the nonlinear wavefront from the leading edge gives valuable information about $\Omega_t^{(0)}$.

We note that for an accelerating aerofoil along a straight line, the linear wavefront from the nose develops fold in the caustic region but the nonlinear wavefront does not fold and has a pair of kinks. For a supersonic aerofoil moving on a highly curved path (curved downwards), the nonlinear wavefront from the upper surface is smooth but that from the lower surface has a pair of kinks.

In Fig. 10.11 a diamond shaped aerofoil, moving from right to left, accelerates from a Mach number 1.2 to 4.0. The figure depicts the boom calculated by the Euler's numerical solution [24]. The leading shock (LS) from the front edge tends to focus

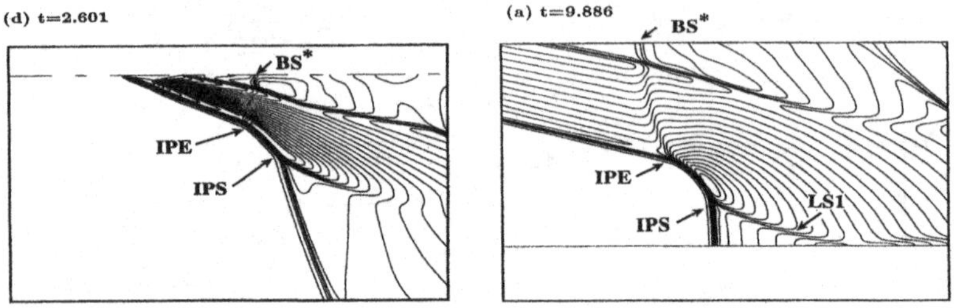

Fig. 9.11 Numerical solutions of Euler equations for the flow over a diamond-shaped projectile moving from right to left with Mach number accelarating from 1.2 to 4. This figure has been reproduced from [9]—Fig. 5.4a, b. with permission from the Indian Academy of Sciences

and develops two kinks (marked by IPE and IPS). Though a property like smoothness of the trailing shock is hard to see in and predict from a numerical result, we note that the trailing shock (from upper shock in the figure) to be smooth.

Chapter 10
3-D WNLRT and SRT: Theory and Applications

The main aim of this chapter is to analyse equations of 3-D WNLRT and those of 3-D SRT and use them to solve some important problems. The results in the chapter are reproduced from [5]. We begin this chapter by writing equations of 3-D WNLRT, which are scattered at various places in earlier chapters.

3-D KCL: We introduced the ray coordinates on a moving surface Ω_t in multi-dimensions in the beginning of Chap. 6. Figure 6.1 shows the geometry of a surface Ω_t and ray coordinates ξ_1 and ξ_2 on it for isotropic evolution of Ω_t. The 3-D KCL system is

$$(g_1 \boldsymbol{u})_t - (m\boldsymbol{n})_{\xi_1} = 0, \quad (g_2 \boldsymbol{v})_t - (m\boldsymbol{n})_{\xi_2} = 0 \tag{10.1}$$

along with the geometric solenoidal conditions

$$(g_2 \boldsymbol{v})_{\xi_1} - (g_1 \boldsymbol{u})_{\xi_2} = 0, \tag{10.2}$$

where $\boldsymbol{u}$ and $\boldsymbol{v}$ are unit tangent vectors along ξ_1 and ξ_2 directions, g_1 and g_2 are metrics associated with ξ_1 and ξ_2 coordinates, respectively, and $\boldsymbol{n}$ is unit normal to Ω_t, given by

$$\boldsymbol{n} = \frac{\boldsymbol{u} \times \boldsymbol{v}}{|\boldsymbol{u} \times \boldsymbol{v}|}. \tag{10.3}$$

The transformation between the (x_1, x_2, x_3)-space and (ξ_1, ξ_2, t)-space is given by (6.4), which gives an expression for the Jacobian as

$$J := \frac{\partial(x_1, x_2, x_3)}{\partial(\xi_1, \xi_2, t)} = g_1 g_2 m \sin\chi, \quad 0 < \chi < \pi, \tag{10.4}$$

where $\chi(\xi_1, \xi_2, t)$ is the angle.[1] between the $\boldsymbol{u}$ and $\boldsymbol{v}$, i.e.

$$\cos\chi = \langle \boldsymbol{u}, \boldsymbol{v} \rangle. \tag{10.5}$$

[1]Printed in bold $\boldsymbol{\chi} = (\chi_1, \chi_2, ..., \chi_d)$ is a ray velocity and not an angle.

P. Prasad, *Propagation of Multidimensional Nonlinear Waves and Kinematical Conservation Laws*, Infosys Science Foundation Series,
https://doi.org/10.1007/978-981-10-7581-0_10

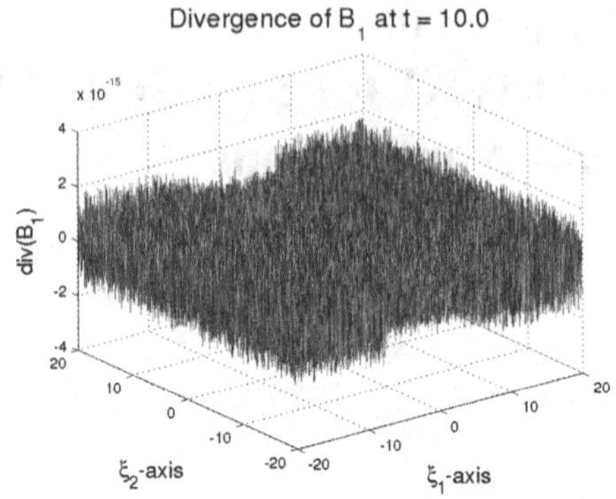

Fig. 10.1 First component of the discrete divergence defined in [5] at $t = 10.0$. The error is of the order of 10^{-15}

We first recollect the Theorem 6.1.1, which implies here that *'for smooth* Ω_t, (10.1) *implies that if* (10.2) *is satisfied at* $t = 0$, *then it is satisfied for all time'*. A constrained transport (CT) type of numerical technique has been developed by Arun in [1] and has been described in Sect. 10.3. In all numerical solutions of SRT reported in this chapter, using the CT scheme, the constraint (10.2) is satisfied to machine accuracy [5] (see Fig. 10.1) even when Ω_t is not smooth due to the appearance of a kink line on Ω_t.

Then, we recollect the Theorem 6.2.1 which implies here that *'for evolution of a smooth* Ω_t, *the* 3-*D ray Eqs.* (6.1) *and* (6.2) *are equivalent to the* 3-*D KCL* (10.2)'.

The closure relation for 3-D WNLRT is (see (7.10)),

$$\left\{(m-1)^2 e^{2(m-1)} g_1 g_2 \sin\chi\right\}_t = 0, \quad 0 < \chi < \pi. \tag{10.6}$$

In notations of this section, Eqs. (6.7) and (6.8) give following differential forms of 3-D KCL system (10.1) and (10.2)

$$g_{1t} = -m\langle \boldsymbol{n}, \boldsymbol{u}_{\xi_1}\rangle, \tag{10.7}$$

$$g_{2t} = -m\langle \boldsymbol{n}, \boldsymbol{v}_{\xi_2}\rangle, \tag{10.8}$$

$$g_1 \boldsymbol{u}_t = m_{\xi_1}\boldsymbol{n} + m\langle \boldsymbol{n}, \boldsymbol{u}_{\xi_1}\rangle \boldsymbol{u} + m\boldsymbol{n}_{\xi_1} \tag{10.9}$$

and

$$g_2 \boldsymbol{v}_t = m_{\xi_2}\boldsymbol{n} + m\langle \boldsymbol{n}, \boldsymbol{v}_{\xi_2}\rangle \boldsymbol{v} + m\boldsymbol{n}_{\xi_2}. \tag{10.10}$$

Now, we write below the differential form of (10.6). In this case we use the equation (10.5) and also use $u_3^2 = 1 - u_1^2 - u_2^2$ and $v_3^2 = 1 - v_1^2 - v_2^2$ to get

$$g_2 g_{1t} + g_1 g_{2t} + g_1 g_2 \cot\chi \left\{-\frac{n_2}{u_3}u_{1t} + \frac{n_1}{u_3}u_{2t} + \frac{n_2}{v_3}v_{1t} - \frac{n_1}{v_3}v_{2t}\right\} + \frac{2 g_1 g_2 m}{m-1} m_t = 0. \tag{10.11}$$

3-D SRT: The KCL equations are the same as above but since we use capital letters for the dependent variables in shock ray theory, we rewrite them again

$$(G_1 \boldsymbol{U})_t - (M\boldsymbol{N})_{\xi_1} = 0, \quad (G_2 \boldsymbol{V})_t - (M\boldsymbol{N})_{\xi_2} = 0 \tag{10.12}$$

along with the geometric solenoidal conditions

$$\nabla_\xi \mathfrak{B}_k \equiv (G_2 \boldsymbol{V})_{\xi_1} + \{-(G_1 \boldsymbol{U})\}_{\xi_2} = 0. \tag{10.13}$$

Closure relations for 3-D SRT are (see Eqs. (7.15) and (7.18))

$$\left\{(M-1)^2 e^{2(M-1)} G_1 G_2 \sin \Psi\right\}_t + 2M(M-1)^2 e^{2(M-1)} G_1 G_2 \mathcal{V} \sin \Psi = 0 \tag{10.14}$$

and

$$\left\{e^{2(M-1)} G_1 G_2 \mathcal{V}^2 \sin \Psi\right\}_t + (M+1) e^{2(M-1)} G_1 G_2 \mathcal{V}^3 \sin \Psi = 0, \tag{10.15}$$

where $\mathcal{V}$ is a measure of the gradient of the flow behind the shock defined in (5.50) and Ψ, $(0 < \Psi < \pi)$ is the angle between the two unit vectors $\boldsymbol{U}$ and $\boldsymbol{V}$.

10.1 Eigen Structure of the WNLRT Equations

As mentioned in Sect. 6.1, out of the six scalar equations in (10.9) and (10.10) only four are linearly independent. We shall now write the Eqs. (10.7)–(10.11) in terms of seven dependent variables $g_1, g_2, u_1, u_2, v_1, v_2$ and m. For this we use $u_3^2 = 1 - u_1^2 - u_2^2$ and $v_3^2 = 1 - v_1^2 - v_2^2$ (see [3] for details). Equations (10.9) and (10.10) give

$$g_1 u_{1t} - n_1 m_{\xi_1} + b_{11}^{(1)} u_{1\xi_1} + b_{12}^{(1)} u_{2\xi_1} + b_{13}^{(1)} v_{1\xi_1} + b_{14}^{(1)} v_{2\xi_1} = 0, \tag{10.16}$$

$$g_1 u_{2t} - n_2 m_{\xi_1} + b_{21}^{(1)} u_{1\xi_1} + b_{22}^{(1)} u_{2\xi_1} + b_{23}^{(1)} v_{1\xi_1} + b_{24}^{(1)} v_{2\xi_1} = 0, \tag{10.17}$$

$$g_2 v_{1t} - n_1 m_{\xi_2} + b_{31}^{(2)} u_{1\xi_2} + b_{32}^{(2)} u_{2\xi_2} + b_{33}^{(2)} v_{1\xi_2} + b_{34}^{(2)} v_{2\xi_2} = 0, \tag{10.18}$$

$$g_2 v_{2t} - n_2 m_{\xi_2} + b_{41}^{(2)} u_{1\xi_2} + b_{42}^{(2)} u_{2\xi_2} + b_{43}^{(2)} v_{1\xi_2} + b_{44}^{(2)} v_{2\xi_2} = 0, \tag{10.19}$$

where the coefficients $b_{ij}^{(1)}$ and $b_{ij}^{(2)}$ are given in the Appendix 1 at the end of this chapter. We write the Eq. (10.11) in the form

$$a_{51} u_{1t} + a_{52} u_{2t} + a_{53} v_{1t} + a_{54} v_{2t} + a_{55} m_t + a_{56} g_{1t} + a_{57} g_{2t} = 0. \tag{10.20}$$

Differential forms of Eqs. (10.7) and (10.8) are

$$g_{1t} + b_{61}^{(1)} u_{1\xi_1} + b_{62}^{(1)} u_{2\xi_1} = 0, \tag{10.21}$$

$$g_{2t} + b_{73}^{(2)} v_{1\xi_2} + b_{74}^{(2)} v_{2\xi_2} = 0. \tag{10.22}$$

A matrix form of these equations for the vector $\boldsymbol{U} = (u_1, u_2, v_1, v_2, m, g_1, g_2)^T$ is

$$A\boldsymbol{U}_t + B^{(1)}\boldsymbol{U}_{\xi_1} + B^{(2)}\boldsymbol{U}_{\xi_2} = 0, \tag{10.23}$$

where

$$A = \begin{bmatrix} g_1 & 0 & 0 & 0 & 0 & 0 & 0 \\ 0 & g_1 & 0 & 0 & 0 & 0 & 0 \\ 0 & 0 & g_2 & 0 & 0 & 0 & 0 \\ 0 & 0 & 0 & g_2 & 0 & 0 & 0 \\ a_{51} & a_{52} & a_{53} & a_{54} & a_{55} & a_{56} & a_{57} \\ 0 & 0 & 0 & 0 & 0 & 1 & 0 \\ 0 & 0 & 0 & 0 & 0 & 0 & 1 \end{bmatrix}, \tag{10.24}$$

$$B^{(1)} = \begin{bmatrix} b_{11}^{(1)} & b_{12}^{(1)} & b_{13}^{(1)} & b_{14}^{(1)} & -n_1 & 0 & 0 \\ b_{21}^{(1)} & b_{22}^{(1)} & b_{23}^{(1)} & b_{24}^{(1)} & -n_2 & 0 & 0 \\ 0 & 0 & 0 & 0 & 0 & 0 & 0 \\ 0 & 0 & 0 & 0 & 0 & 0 & 0 \\ 0 & 0 & 0 & 0 & 0 & 0 & 0 \\ b_{61}^{(1)} & b_{62}^{(1)} & 0 & 0 & 0 & 0 & 0 \\ 0 & 0 & 0 & 0 & 0 & 0 & 0 \end{bmatrix}, \tag{10.25}$$

$$B^{(2)} = \begin{bmatrix} 0 & 0 & 0 & 0 & 0 & 0 & 0 \\ 0 & 0 & 0 & 0 & 0 & 0 & 0 \\ b_{31}^{(2)} & b_{32}^{(2)} & b_{33}^{(2)} & b_{34}^{(2)} & -n_1 & 0 & 0 \\ b_{41}^{(2)} & b_{42}^{(2)} & b_{43}^{(2)} & b_{44}^{(2)} & -n_2 & 0 & 0 \\ 0 & 0 & 0 & 0 & 0 & 0 & 0 \\ 0 & 0 & 0 & 0 & 0 & 0 & 0 \\ 0 & 0 & b_{73}^{(2)} & b_{74}^{(2)} & 0 & 0 & 0 \end{bmatrix}. \tag{10.26}$$

This system of equations is quite complex and we could not calculate the eigenvalues $\lambda_i,\ i = 1, \cdots, 7$ of the system by explicitly solving the characteristic equation

$$M_{7,7}(\lambda) := \det\left(-\lambda A + e_1 B^{(1)} + e_2 B^{(2)}\right) = 0, \tag{10.27}$$

where $(e_1, e_2) \in \mathbb{R}^2 \backslash (0, 0)$. Therefore, we used quite a few novel methods to discuss the eigenvalues and their multiplicity in [3, 4]. We present here the simplest of these. Note that on an arbitrary moving surface Ω_t, we may choose a pair of orthogonal surface coordinates ξ_1 and ξ_2 at a fixed time t, say at $t = 0$ but later when the surface evolves according to the Eqs. (10.1), (10.2) and (10.6), the coordinates ξ_1 and ξ_2 need not remain orthogonal. However, we mention a result:

Remark 10.1.1 Let $P_0(\boldsymbol{x}_0)$ be a given point on Ω_t at a fixed time t. Then, there exist two one-parameter families of smooth curves on Ω_t such that the unit vectors $\boldsymbol{u}_0$ and $\boldsymbol{v}_0$ along the members of the curves through the **chosen** point $P_0(\boldsymbol{x}_0)$ can have any two arbitrary directions and the metrics g_{10} and g_{20} at this point can have any two positive values.

Now, we choose a coordinate system (η_1, η_2, t) such that the tangent vectors $\boldsymbol{u}'$ and $\boldsymbol{v}'$ along the coordinates are orthogonal at some given point P_0 on Ω_t. Then, at P_0, $\langle \boldsymbol{u}', \boldsymbol{v}' \rangle = \cos\frac{\pi}{2} = 0$. Let the metrics associated with this system be denoted by g_1' and g_2'. In the coordinate system (η_1, η_2, t) the characteristic equation (10.27) at P_0, namely

$$\det\left(-\lambda' A + e_1' B^{(1)} + e_2' B^{(2)}\right) = 0, \quad (e_1', e_2') \in \mathbb{R}^2 \backslash (0, 0) \tag{10.28}$$

of the KCL-based WNLRT simplifies very much and becomes

$$\det\begin{bmatrix} -\lambda' g_1' & 0 & \frac{mu_2'v_1'}{v_3'}e_1' & \frac{mu_1'v_1'}{v_3'}e_1' & -n_1e_1' & 0 & 0 \\ 0 & -\lambda' g_1' & \frac{mu_2'v_2'}{v_3'}e_1' & \frac{mu_1'v_2'}{v_3'} & -n_2e_1' & 0 & 0 \\ -\frac{mu_1'v_2'}{u_3'}e_2' & \frac{mu_1'v_1'}{u_3'}e_2' & -\lambda' g_2' & 0 & -n_1e_2' & 0 & 0 \\ -\frac{mu_2'v_2'}{u_3'}e_2' & \frac{mu_2'v_1'}{u_3'}e_2' & 0 & -\lambda' g_2' & -n_2e_2' & 0 & 0 \\ 0 & 0 & 0 & 0 & -\lambda'\frac{2m}{m-1}g_1'g_2' & -\lambda' g_2' & -\lambda' g_1' \\ -\frac{mv_2'}{u_3'}e_1' & \frac{mv_1'}{u_3'}e_1' & 0 & 0 & 0 & -\lambda' & 0 \\ 0 & 0 & \frac{mu_2'}{v_3'}e_2' & -\frac{mu_1'}{v_3'}e_2' & 0 & 0 & -\lambda' \end{bmatrix} = 0. \tag{10.29}$$

Some calculations lead to the following eigenvalues:

$$\lambda_{1,2}' = \pm\left\{\frac{m-1}{2}\left(\frac{e_1'^2}{g_1'^2} + \frac{e_2'^2}{g_2'^2}\right)\right\}^{1/2}, \quad \lambda_3' = \lambda_4' = \lambda_5' = \lambda_6' = \lambda_7' = 0. \tag{10.30}$$

It is also found that the number of independent eigenvectors corresponding to the multiple eigenvalue 0 is 4 resulting in the loss of hyperbolicity of the system for $m > 1$.

To get the eigenvalues at P_0 in the general surface coordinates (ξ_1, ξ_2) on Ω_t, we need to find the linear transformation of the tangent vectors $(\boldsymbol{u}', \boldsymbol{v}')$ associated with the coordinates (η_1, η_2) to the tangent vectors $(\boldsymbol{u}, \boldsymbol{v})$ associated with the coordinates (ξ_1, ξ_2) at P_0. Let this transformation in the tangent plane at P_0 of Ω_t be given by

$$\boldsymbol{u}' = \gamma_1\boldsymbol{u} + \delta_1\boldsymbol{v}, \quad \boldsymbol{v}' = \gamma_2\boldsymbol{u} + \delta_2\boldsymbol{v}. \tag{10.31}$$

The following theorem in [3] helps us to get the eigenvalues of the KCL-based WNLRT in any ray coordinate system (ξ_1, ξ_2, t) from the expressions for λ_1' and λ_2' in (10.30).

Theorem 10.1.2 *Let λ' be an expression of an eigenvalue of the KCL-based WNLRT at a point P_0 in ray coordinates (η_1, η_2, t) in terms of e_1'/g_1' and e_2'/g_2'. Then, the expression for the same eigenvalue (now denoted by λ) at P_0 in ray coordinates (ξ_1, ξ_2, t) in terms of e_1/g_1 and e_2/g_2 can be obtained by substituting the following expressions for e_1'/g_1' and e_2'/g_2':*

$$\lambda' = \lambda, \quad \frac{e_1'}{g_1'} = \gamma_1 \frac{e_1}{g_1} + \delta_1 \frac{e_2}{g_2}, \quad \frac{e_2'}{g_2'} = \gamma_2 \frac{e_1}{g_1} + \delta_2 \frac{e_2}{g_2}. \tag{10.32}$$

Proof The proof is very simple [3]. First, we use the transformation of derivations

$$\frac{1}{g_1'} \frac{\partial}{\partial \eta_1} = \gamma_1 \frac{1}{g_1} \frac{\partial}{\partial \xi_1} + \delta_1 \frac{1}{g_2} \frac{\partial}{\partial \xi_2}, \tag{10.33}$$

$$\frac{1}{g_2'} \frac{\partial}{\partial \eta_2} = \gamma_2 \frac{1}{g_1} \frac{\partial}{\partial \xi_1} + \delta_2 \frac{1}{g_2} \frac{\partial}{\partial \xi_2} \tag{10.34}$$

in the equation

$$A \frac{\partial \boldsymbol{U}'}{\partial t} + B'^{(1)} \frac{\partial \boldsymbol{U}'}{\partial \eta_1} + B'^{(2)} \frac{\partial \boldsymbol{U}'}{\partial \eta_2} = 0 \tag{10.35}$$

and compare the coefficients of the transformed system with those of the Eq. (10.23). This gives

$$g_1 B^{(1)} = \gamma_1 g_1' B'^{(1)} + \gamma_2 g_2' B'^{(2)} \quad \textit{and} \quad g_2 B^{(2)} = \delta_1 g_1' B'^{(1)} + \delta_2 g_2' B'^{(2)}.$$

We use these to transform characteristic equation (10.27) of the system in (ξ_1, ξ_2, t) and compare the result with (10.28) to get the result (10.32). ■

Now, we proceed to derive the eigenvalues of the system (10.23). Given the surface coordinates (ξ_1, ξ_2) at a point P_0 on Ω_t with unit tangent vectors $\boldsymbol{u}, \boldsymbol{v}$ and $\langle \boldsymbol{u}, \boldsymbol{v} \rangle = \cos \chi$, we choose an orthogonal coordinate at P_0 with unit vectors $\boldsymbol{u}', \boldsymbol{v}'$ as follows:

$$\boldsymbol{u}' = \boldsymbol{u}, \quad \boldsymbol{v}' = \gamma_2 \boldsymbol{u} + \delta_2 \boldsymbol{v}, \tag{10.36}$$

where $\langle \boldsymbol{u}', \boldsymbol{v}' \rangle = 0$ and $|\boldsymbol{v}'| = 1$ so that

$$\gamma_2 = -\cot \chi, \quad \delta_2 = \csc \chi, \tag{10.37}$$

where csc stands for cosec. The sign of γ_2 and δ_2 are chosen so that vectors $\boldsymbol{u}', \boldsymbol{v}'$ and $\boldsymbol{n}' = \boldsymbol{n}$ form a right- handed system, similar to the three vectors $\boldsymbol{u}, \boldsymbol{v}$ and $\boldsymbol{n}$.

Substituting $\gamma_1 = 1$, $\delta_1 = 0$ and the above values of γ_2 and δ_2 in the expression (10.32), we get

$$\frac{e_1'}{g_1'} = \frac{e_1}{g_1}, \quad \frac{e_2'}{g_2'} = -\cot\chi\frac{e_1}{g_1} + \csc\chi\frac{e_2}{g_2}. \tag{10.38}$$

Substituting these in (10.30), we get the eigenvalues of (10.23) as

$$\lambda_{1,2} = \pm\left\{\frac{m-1}{2\sin^2\chi}\left(\frac{e_1^2}{g_1^2} - \frac{2e_1e_2}{g_1g_2}\cos\chi + \frac{e_2^2}{g_2^2}\right)\right\}^{1/2}, \quad \lambda_3 = \lambda_4 = \lambda_5 = \lambda_6 = \lambda_7 = 0. \tag{10.39}$$

We note that $\frac{e_1}{g_1}$ and $\frac{e_2}{g_2}$ cannot be taken to be zero simultaneously. Since $\boldsymbol{u}$ and $\boldsymbol{v}$ also cannot be taken parallel, $\sin\chi \neq 0$. From the expression for λ_1, it follows that λ_1 is real and positive for $m - 1 > 0$ and purely imaginary for $m - 1 < 0$.

Eigenspace corresponding to the eigenvalue $\lambda = 0$: As reported in [3], extensive numerical computation points out that for the eigenvalue 0, the eigenspace is 4-D. In fact, the rank of the pencil matrix of the system (10.23) for the eigenvalue $\lambda = 0$, i.e. the rank of $e_1B^{(1)} + e_2B^{(2)}$ will be the same as the rank of $e_1'B'^{(1)} + e_2'B'^{(2)}$ when the relations $g_1B^{(1)} = \gamma_1g_1'B'^{(1)} + \gamma_2g_2'B'^{(2)}$ *and* $g_2B^{(2)} = \delta_1g_1'B'^{(1)} + \delta_2g_2'B'^{(2)}$ are valid and hence the rank of $e_1B^{(1)} + e_2B^{(2)}$ would be 3. Thus, the number of linearly independent eigenvectors corresponding to $\lambda = 0$ is only $7 - 3 = 4$.

We have now proved the main theorem.

Theorem 10.1.3 *The system (10.23) has 7 eigenvalues $\lambda_1, \lambda_2 (= -\lambda_1), \lambda_3 = \lambda_4 = \ldots = \lambda_7 = 0$, where λ_1 and λ_2 are real for $m > 1$ and purely imaginary for $m < 1$. Further, the dimension of the eigenspace corresponding to the multiple eigenvalue 0 is 4.*

10.2 Eigen Structure of the SRT Equations

In order to discuss the eigenvalues and eigen structure of the system formed by (10.12)–(10.15), we need its explicit form as a system of partial differential equations. Introducing a vector of primitive variables via $V = (U_1, U_2, V_1, V_2, M, G_1, G_2, \mathcal{V})^T$, we can derive a quasilinear form of (10.12)–(10.15) as

$$\tilde{A}V_t + \tilde{B}^{(1)}V_{\xi_1} + \tilde{B}^{(2)}V_{\xi_2} = \tilde{C}, \tag{10.40}$$

where the expressions for the matrices can be derived following the procedure of the Sect. 10.1 and are explicitly given in the appendix of [5] reproduced in Appendix 2 at the end of this Chapter. It is interesting to note that the matrices $\tilde{A}$, $\tilde{B}^{(1)}$ and $\tilde{B}^{(2)}$ admit the following block structure:

$$\tilde{A} = \begin{bmatrix} A & O_{7,1} \\ R_{1,7} & \frac{2G_1G_2}{\mathcal{V}} \end{bmatrix}, \quad \tilde{B}^{(1)} = \begin{bmatrix} B^{(1)} & O_{7,1} \\ O_{1,7} & 0 \end{bmatrix}, \quad \tilde{B}^{(2)} = \begin{bmatrix} B^{(2)} & O_{7,1} \\ O_{1,7} & 0 \end{bmatrix}. \tag{10.41}$$

Here, A, $B^{(1)}$ and $B^{(2)}$ are the corresponding Jacobian matrices of 3-D WNLRT given by (10.24)–(10.26) (but variables replaced appropriately by upper case letters as required for SRT), O denotes a matrix with entries 0 and $R_{1,7}$ is a row matrix. The characteristic equation of (10.40) is given by

$$\det \tilde{M}_{8,8}(\lambda) \equiv \det\left(e_1\tilde{B}^{(1)} + e_2\tilde{B}^{(2)} - \lambda\tilde{A}\right) = 0. \tag{10.42}$$

Using the block structure of the matrices in (10.41), we can obtain

$$\tilde{M}_{8,8}(\lambda) = \begin{bmatrix} M_{7,7}(\lambda) & O_{7,1} \\ -\lambda R_{1,7} & -\lambda\frac{2G_1G_2}{\mathcal{V}} \end{bmatrix}, \tag{10.43}$$

where $M_{7,7}(\lambda)$ is the matrix pencil of the 3-D WNLRT as defined in (10.27), but the variables replaced appropriately by upper case letters as required for SRT. Therefore, the characteristic equation (10.42) simplifies to

$$\lambda \det M_{7,7}(\lambda) = 0. \tag{10.44}$$

Using the results of the Sect. 10.1, it is easy to find the roots of the polynomial equation $\det M_{7,7}(\lambda) = 0$ in λ and the nullvectors of $M_{7,7}$ corresponding to these roots. The roots of (10.44) can be obtained as $\lambda_1, \lambda_2 (= -\lambda_1), \lambda_3 = \cdots = \lambda_8 = 0$, where

$$\lambda_1 = \left\{\frac{M-1}{2\sin^2\Psi}\left(\frac{E_1^2}{G_1^2} - \frac{2E_1E_2}{G_1G_2}\cos\Psi + \frac{E_2^2}{G_2^2}\right)\right\}^{\frac{1}{2}} \tag{10.45}$$

and $(E_1, E_2) \in \mathbb{R}^2$ with $E_1^2 + E_2^2 = 1$. Let us now consider the matrix $\tilde{M}_{8,8}(0)$ for the multiple eigenvalue $\lambda = 0$. Clearly, the rank of the matrix $\tilde{M}_{8,8}(0)$ is same as that of $M_{7,7}(0)$ in (10.27). It has been shown in Sect. 10.1 that the rank of $M_{7,7}(0)$ is three. Thus, the dimension of the nullspace of $\tilde{M}_{8,8}(0)$ is $8 - 3 = 5$. This proves that the dimension of the eigenspace corresponding to $\lambda = 0$ of multiplicity 6 is only 5. We summarize these results as a theorem.

Theorem 10.2.1 *The system (10.40) has eight eigenvalues $\lambda_1, \lambda_2\ (= -\lambda_1), \lambda_3 = \lambda_4 = \cdots = \lambda_8 = 0$, where λ_1 and λ_2 are real for $M > 1$ and purely imaginary for $M < 1$. Further, the dimension of the eigenspace corresponding to the multiple eigenvalue 0 is 5.*

Remark 10.2.2 It is important to note that $m < 1$ and $M < 1$ are not physically unrealistic for a nonlinear wavefront and a shock front respectively. The signature of a sonic boom produced by a convex and smooth upper surface an aerofoil with a sharp leading edge (see Fig. 9.8-2) consists of a leading shock followed by a continuous flow in which the pressure decreases and the continuous flow terminates in a trailing shock. In the forward part of this continuous flow, the pressure is greater than that in the ambient medium ahead of the leading shock and in the rear part it is less than that behind the trailing shock.[2] The shock velocity of the trailing weak shock is half of the sound velocity behind it, say a_0 and the sound speed ahead of it, which is less than a_0. Thus Mach number M of the trailing shock is less than 1.

10.3 Numerical Approximation of the SRT Equations

As a consequence of Theorem 10.2.1, we infer that the system of conservation laws (10.40) is only weakly hyperbolic for $M > 1$; hence, an initial value problem may not be well posed as it is in the case of strong hyperbolic system. In addition, weakly hyperbolic systems are likely to be more sensitive than regular hyperbolic systems also from a computational point of view. Numerical as well as theoretical analysis indicates that their solution may not belong to the BV spaces and can only be measure valued. Despite such theoretical difficulties, in [1, 2, 5], we have been able to develop accurate and efficient numerical approximations of the analogous weakly hyperbolic system of 3-D WNLRT using, simple but robust, central schemes. Since the conservation laws of 3-D SRT and those of 3-D WNLRT, viz. (10.1) and (10.6), are structurally similar, we do not intend to present the details of the numerical approximation and the refer the reader to [1] for more details.

We discretize the system (10.40) using the cell integral averages $\overline{W}_{i,j}$ of the conservative variable W, taken over square mesh cells. From the given cell averages $\overline{W}^n_{i,j}$ at time t^n, we reconstruct a piecewise linear interpolant using the standard MUSCL-type procedures. In order to obtain the discrete slopes in the ξ_1- and ξ_2-directions, we employ a central weighted essentially non-oscillatory limiter [25]. The piecewise linear reconstruction enables us to compute the cell interface values of the conserved variable W.

The starting point for the construction of numerical scheme is a semi-discrete discretization of (10.40), given by

$$\frac{d\overline{W}_{i,j}}{dt} = -\frac{\mathcal{F}_{1i+\frac{1}{2},j} - \mathcal{F}_{1i-\frac{1}{2},j}}{h_1} - \frac{\mathcal{F}_{2i,j+\frac{1}{2}} - \mathcal{F}_{2i,j-\frac{1}{2}}}{h_2} + S(\overline{W}_{i,j}), \tag{10.46}$$

where the quantities $\mathcal{F}_{i+1/2,j}$ and $\mathcal{F}_{i,j+1/2}$ are, respectively, the numerical fluxes at the cell interfaces $(i+1/2, j)$ and $(i, j+1/2)$. We employ the high- resolution flux given in [27], for these interface fluxes, e.g. at a right-hand vertical edge

$$\mathcal{F}_{1i+\frac{1}{2},j}\left(W^R_{i,j}, W^L_{i+1,j}\right) = \frac{1}{2}\left(F_1\left(W^L_{i+1,j}\right) + F_1\left(W^R_{i,j}\right)\right) - \frac{a_{i+\frac{1}{2},j}}{2}\left(W^L_{i+1,j} - W^R_{i,j}\right), \tag{10.47}$$

where $W^{L(R)}_{i,j}$ denote, respectively, the left- and right- interpolated states at the interface $(i+1/2, j)$. The expression for the flux $\mathcal{F}_{i,j+1/2}$ at an upper horizontal edge is analogous. In the flux formula (10.47), the term $a_{i+1/2,j}$ denotes the local speed of

[2]This continuous flow has a sonic line starting from a point P_G on the aerofoil, see Fig. 9.8-2.

propagation at cell interfaces, given by

$$a_{i+\frac{1}{2},j} := \max\left\{\rho\left(\frac{\partial F_1}{\partial W}\left(W^R_{i,j}\right)\right), \rho\left(\frac{\partial F_1}{\partial W}\left(W^L_{i+1,j}\right)\right)\right\}, \tag{10.48}$$

where $\rho(A) := max_i|\lambda_i(A)|$, with $\lambda_i(A)$ being the eigenvalues of the matrix A.

To improve the temporal accuracy and to gain second-order accuracy in time, we use a TVD Runge–Kutta scheme [53] to numerically integrate the system of ordinary differential equations in (10.46). Denoting the right-hand side of (10.46) by $\mathcal{L}_{i,j}(W)$, the second order Runge–Kutta scheme updates W through the following two stages:

$$W^{(1)}_{i,j} = \overline{W}^n_{i,j} + \Delta t\mathcal{L}_{i,j}\left(\overline{W}^n\right), \tag{10.49}$$

$$\overline{W}^{n+1}_{i,j} = \frac{1}{2}\overline{W}^n_{i,j} + \frac{1}{2}W^{(1)}_{i,j} + \frac{1}{2}\Delta t\mathcal{L}_{i,j}\left(W^{(1)}\right). \tag{10.50}$$

It is to be noted that any consistent numerical solution of the 3-D SRT system (10.12), (10.14) and (10.15) also has to satisfy the geometric solenoidal constraint (10.13) at any time. Note that this constraint is an involution for the 3-D KCL (10.12), i.e. once fulfilled at the initial data it is fulfilled at all times. Since the physically exact solution has this feature, a numerical solution should also possess some discrete sense. Hence, in the numerical approximation of the analogous 3-D WNLRT system in [1], a constrained transport (CT) algorithm [18] was built into the central finite volume to enforce the geometric solenoidal constraint. In what follows, we briefly review this CT strategy and refer to [1] for more details on its implementation.

The geometric solenoidal constraint (10.13) guarantees the existence of three potential functions $\mathbb{A}_k$, $k = 1, 2, 3$, such that

$$G_1U_k = \mathbb{A}_{k\xi_1}, \; G_2V_k = \mathbb{A}_{k\xi_2}. \tag{10.51}$$

Using (10.51) in the 3-D KCL system (10.12) yields the evolution equations

$$\mathbb{A}_{kt} = MN_k. \tag{10.52}$$

In the CT method, we store the three potentials $\mathbb{A}_k$ at the centres of a staggered grid. With the aid of these potentials, we redefine the values of the vectors $G_1\boldsymbol{U}$ and $G_2\boldsymbol{V}$ at the cell edges, which are treated as length averaged quantities. In this way, $G_2\boldsymbol{V}$ is collocated at $(i + 1/2, j)$, whereas $G_1\boldsymbol{U}$ is collocated at $(i, j + 1/2)$. The definitions of these collocated values are obtained by simply discretizing the derivatives in (10.51) using central differences, i.e.

$$[G_1U_k]_{i,j+\frac{1}{2}} = \frac{1}{h_1}\left(\mathbb{A}_{k\,i+\frac{1}{2},j+\frac{1}{2}} - \mathbb{A}_{k\,i-\frac{1}{2},j+\frac{1}{2}}\right), \tag{10.53}$$

$$[G_2 V_k]_{i+\frac{1}{2},j} = \frac{1}{h_2}\left(\mathbb{A}_{k i+\frac{1}{2},j+\frac{1}{2}} - \mathbb{A}_{k i+\frac{1}{2},j-\frac{1}{2}}\right). \tag{10.54}$$

With the above collocated values, we calculate

$$\begin{aligned}(G_2 V_k)_{\xi_1} - (G_1 U_k)_{\xi_2}\big|_{i,j} &= \frac{1}{h_1}\left([G_2 V_k]_{i+\frac{1}{2},j} - [G_2 V_k]_{i-\frac{1}{2},j}\right) \\ &\quad - \frac{1}{h_2}\left([G_1 U_k]_{i,j+\frac{1}{2}} - [G_1 U_k]_{i,j-\frac{1}{2}}\right).\end{aligned} \tag{10.55}$$

Using (10.53) and (10.54), we can easily see that the right-hand side of (10.55) vanishes due to perfect cancellation. Hence, in this way, we have devised a method to enforce the geometric solenoidal constraint at the cell centres of the finite volumes.

It remains to be specified how to compute the values of the potentials $\mathbb{A}_k$ on the staggered grids. Note that integrating (10.52) over a staggered grid yields the following update formula:

$$\frac{d}{dt}\mathbb{A}_{k i+\frac{1}{2},j+\frac{1}{2}} = [MN_k]_{i+\frac{1}{2},j+\frac{1}{2}}. \tag{10.56}$$

In order to compute the expression on the right-hand side we use a simple averaging, i.e.

$$[MN_k]_{i+\frac{1}{2},j+\frac{1}{2}} = \frac{1}{4}\left([MN_k]_{i+\frac{1}{2},j} + [MN_k]_{i,j+\frac{1}{2}} + [MN_k]_{i+\frac{1}{2},j+1} + [MN_k]_{i+1,j+\frac{1}{2}}\right), \tag{10.57}$$

where the values of MN_k on the cell edges are obtained from the numerical fluxes $\mathcal{F}_1$ and $\mathcal{F}_2$ of the finite volume scheme. The resulting ordinary differential equations in (10.56) are integrated using the same TVD Runge–Kutta method in (10.49) and (10.50). At the beginning of the next time step, the cell-centred values of $G_1 U_k$ and $G_2 V_k$ are calculated by interpolation

$$[G_1 U_k]_{i,j} = \frac{1}{2}\left([G_1 U_k]_{i,j+\frac{1}{2}} + [G_1 U_k]_{i,j-\frac{1}{2}}\right), \tag{10.58}$$

$$[G_2 V_k]_{i,j} = \frac{1}{2}\left([G_2 V_k]_{i+\frac{1}{2},j} + [G_2 V_k]_{i-\frac{1}{2},j}\right). \tag{10.59}$$

Thus, we have devised a way to compute the values of $G_1 \boldsymbol{U}$ and $G_2 \boldsymbol{V}$ so that the discrete divergence of the three vectors $\mathfrak{B}_k$ (defined in (10.13)) equals zero. An algorithm, in which the solution of the 3-D SRT equations calculated using the central scheme (10.46) is corrected accordingly in each time step, can be written in the following form:

1. The system of balance equations (10.12–10.15) is solved using the central scheme, giving the values of the conserved variable

$$W = \left(G_1 U, G_2 V, (M-1)^2 e^{2(M-1)} G_1 G_2 \sin\Psi, e^{2(M-1)} G_1 G_2 \mathcal{V}^2 \sin\Psi\right)^T. \tag{10.60}$$

The vectors $\mathfrak{B}_k$ are not yet divergence free, it will be corrected in the next steps.
2. The potentials $\mathbb{A}_1, \mathbb{A}_2, \mathbb{A}_3$ are updated by solving (10.56).
3. The spatial derivatives of $\mathbb{A}_1, \mathbb{A}_2, \mathbb{A}_3$ are calculated and the corrected values of $g_1 U$ and $g_2 V$ are obtained using (10.53) and (10.54).
4. The values of $G_1 U$ and $G_2 V$ on the cell edges are averaged, to get the values at the cell centres of the original grid

$$[G_1 U]^{n+1}_{i,j} = \frac{1}{2}\left([G_1 U]^{n+1}_{i,j-\frac{1}{2}} + [G_1 U]^{n+1}_{i,j+\frac{1}{2}}\right), \tag{10.61}$$

$$[G_2 V]^{n+1}_{i,j} = \frac{1}{2}\left([G_2 V]^{n+1}_{i-\frac{1}{2},j} + [G_2 V]^{n+1}_{i+\frac{1}{2},j}\right). \tag{10.62}$$

5. To get the successive positions of the shock front Ω_t, we numerically integrate the first part, viz. (6.2), of the shock ray equations, i.e.

$$\frac{d}{dt} X_{i,j}(t) = M_{i,j}(t) N_{i,j}(t) \tag{10.63}$$

using the two-stage Runge–Kutta method.

10.3.1 Formulation of Initial Data and Construction of Shock Front at $t > 1$

There is an extensive discussion in [1, 3] on the formulation of the initial data and the implementation of boundary conditions needed for numerical solution for KCL system. We present here briefly.

Let the initial position of a weak shock front Ω_t be given in a parametric form as follows:

$$\Omega_0 : \; x = x_0(\xi_1, \xi_2). \tag{10.64}$$

Given the representation (10.64), the initial values of the metrics G_1 and G_2 and unit tangent vectors U and V can be taken as

$$G_{10} = |x_{0\xi_1}|, \; G_{20} = |x_{0\xi_2}|, \tag{10.65}$$

$$U_0 = \frac{x_{0\xi_1}}{|x_{0\xi_1}|}, \; V_0 = \frac{x_{0\xi_2}}{|x_{0\xi_2}|}. \tag{10.66}$$

Note that

$$(G_{20}V_0)_{\xi_1} - (G_{10}U_0)_{\xi_2} = x_{0\xi_2\xi_1} - x_{0\xi_1\xi_2} = 0. \tag{10.67}$$

Thus, the geometric solenoidal constraint (10.13) is satisfied by the initial values of G_1U and G_2V. The unit normal N_0 of Ω_0 can be obtained using

$$N_0 = \frac{U_0 \times V_0}{|U_0 \times V_0|}. \tag{10.68}$$

Let us assume that the distribution of the shock front velocity M on Ω_0 be given as

$$M = M_0(\xi_1, \xi_2). \tag{10.69}$$

Note: The initial value of the gradient $\mathcal{V}$ of the gas density at the shock front is usually determined by the source which produces the shock (see for example (9.19)). But, in this chapter in all numerical computations we shall choose it a constant positive value, say

$$\mathcal{V}_0 = 0.2. \tag{10.70}$$

$\mathcal{V}_0$ was chosen to be 0.1 for the results in Fig. 9.4.

10.4 Applications of 3-D SRT

A lot of numerical results are available for the propagation of 3-D weakly nonlinear wavefronts and shock fronts in [1–3, 5]. We present here some interesting cases from [5].

10.4.1 Corrugational Stability and Interaction of Kink Lines

The corrugational stability of a front is defined and has been explained; and some results of corrugation of 2-D fronts has been briefly presented in Sect. 9.4. The extensive numerical simulations by Monica and Prasad [32], using 2-D SRT, clearly show that a 2-D shock front is corrugationally stable. In Fig. 9.4 continuous lines are successive positions of the 2-D initially sinusoidal shock front, the broken lines are the rays and dots are the kinks. The shock has become almost a straight line much before $t = 40$. Similarly, the results of numerical experiments with 3-D WNLRT, reported in [1, 2], show that a 3-D nonlinear wavefront is also corrugationally stable.

Our aim here is to verify the corrugational stability of a 3-D shock front, evolving according to 3-D SRT, and describe a very beautiful aspect of interaction of kink lines in more detail.

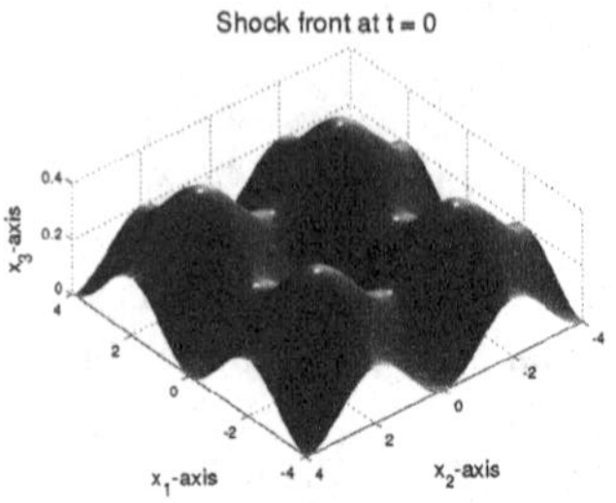

Fig. 10.2 Initial shock front in the shape of a smooth periodic pulse

The corrugational stability is a result of the genuine nonlinearity in the characteristic fields corresponding to the two non-zero eigenvalues of the system (10.12)–(10.15). The shocks in the (ξ_1, ξ_2, t)-coordinates, which are mapped onto kinks, cause dissipation of the kinetic energy. Notice that the energy transport equation (10.6) of 3-D WNLRT is homogeneous, whereas the corresponding equation (10.14) of 3-D SRT has a source term. In the case of a nonlinear wavefront, the value of $m - 1$ converges to the mean value of $m_0 - 1$. For a shock, when $\mathcal{V} > 0$, the value of $M - 1$ decreases to zero. This is typical result of a plane shock in gas dynamics, which can also be seen[3] from the 1-D model equation $u_t + (u^2/2)_x = 0$. Thus, the value of $M-1$, in addition to approaching a constant value, decays. The result is that the perturbations in the shape of a periodic shock front not only disappear, leading to corrugational stability, but also the front velocity M approaches the linear front velocity $M = 1$. In order to verify the corrugational stability, we consider here the initial shock front Ω_0 to be of a periodic shape in x_1- and x_2-directions

$$\Omega_0\colon \; x_3 = \kappa\left(2 - \cos\left(\frac{\pi x_1}{a}\right) - \cos\left(\frac{\pi x_2}{b}\right)\right) \tag{10.71}$$

with the constants $\kappa = 0.1, a = b = 2$. In Fig. 10.2 we give the plot of the initial shock front Ω_0, which is a smooth pulse without any kink lines. The initial front Ω_0 can be thought of modelling a smooth perturbation of a plane front. We prescribe a constant initial velocity $M_0 = 1.2$ everywhere on the front given in (10.71). Though the initial shock front is smooth, as the time evolves, a number of kink lines appear in each period. Now, we describe the very interesting process of interaction of these kink lines. We proceed to do it with a number of plots of the shock front at different instances.

[3]For the scalar conservation law $u_t + (\frac{1}{2}u^2)_x = 0$, we first note the transport equation $\frac{du}{dt} \equiv (\frac{\partial}{\partial t} + u\frac{\partial}{\partial x})u = 0$ of a nonlinear wavefront (see Sect. 3.5). According to this the amplitude $m = u$ remains constant $= m_0$. Then, we refer to the transport equation (5.24) for a shock with $u_r =$ constant, $M - 1 = \frac{u_0 - u_r}{u_r} \equiv \frac{u_\ell - u_r}{u_r}$ and $v_1 = \mathcal{V}$. This gives $\frac{d(M-1)}{dt} = \{\frac{\partial}{\partial t} + \frac{1}{2}(M+1)\frac{\partial}{\partial x}\}(M-1) = -\frac{1}{2}(M-1)\mathcal{V}$. When $\mathcal{V} > 0$, $M - 1 \to 0+$ *as* $t \to \infty$.

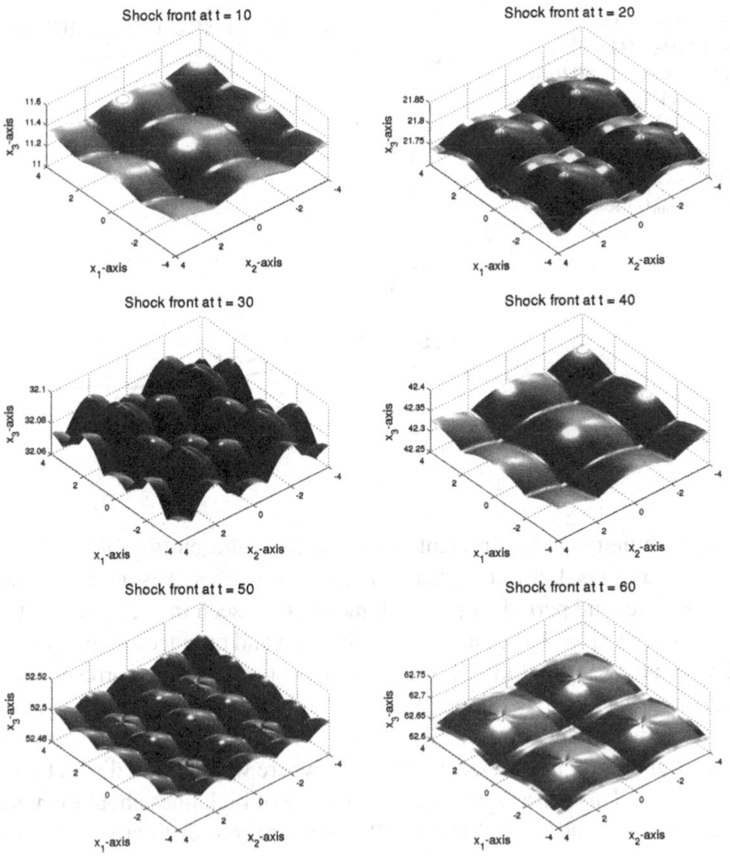

Fig. 10.3 Shock front Ω_t starting initially in a periodic shape with $M_0 = 1.2$ and $\mathcal{V}_0 = 0.2$. The shock front develops a complex pattern of kinks and ultimately becomes planar

In Fig. 10.3, we give the surface plots of the shock front Ω_t at times $t = 10, 20, 30, 40, 50, 60$ in two periods in each of x_1- and x_2-directions. As mentioned above, the initial shock front is smooth, with no kink lines. The front Ω_t moves up in the x_3-direction and develops several kink lines. Four kink lines parallel to x_1-axis and four parallel to x_2-axis can be seen in the figures on the shock front at times $t \geq 10$. These kink lines are formed at a time before $t = 10$, say about $t = 2$. This can easily be observed from the Fig. 10.4 showing[4] the maximum and minimum values, $M_{\max}(t)$ and $M_{\min}(t)$, versus t, where $M_{\max}(t)$ and $M_{\min}(t)$ the maximum and minimum values of M, respectively, taken over (ξ_1, ξ_2) at any time t. We can observe a significant increase in the maximum value of M near $t = 2$ and it also jumps after the interaction of kink lines at later times.

[4] M increases on some parts of Ω_t due to convergence of rays. As soon as the kink lines are formed, the rays tend to diverge as seen in Figs. 9.2B and 9.4, then M starts decreasing.

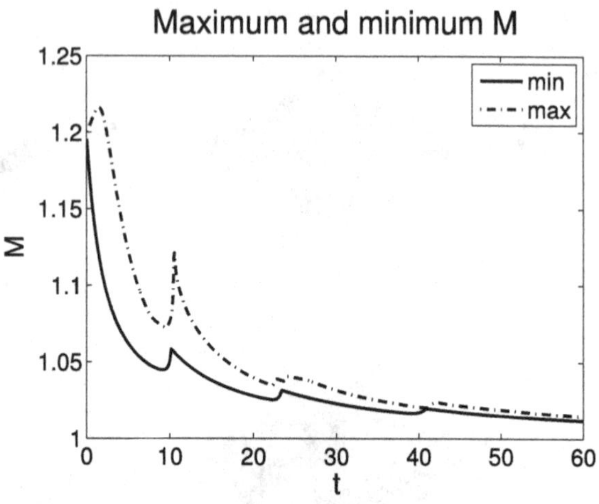

Fig. 10.4 Variation of $M_{\max}(t)$ and $M_{\min}(t)$ with time from $t = 0$ to $t = 80$ for a periodic shock front $x_3 = \kappa\left(2 - \cos\left(\frac{\pi x_1}{2}\right) - \cos\left(\frac{\pi x_2}{2}\right)\right)$. Both $M_{max}(t)$ and $M_{min}(t)$ tend to one and the difference $M_{\max}(t) - M_{\min}(t)$ tends to zero as $t \to \infty$

Due to symmetry, it is sufficient if we describe the motion of the kink lines parallel to the x_1-axis. Let us designate the part $-4 \le x_1 \le 0$ as the first period and $0 \le x_1 \le 4$ as second period. A pair of kink lines are seen in each period at $t = 10$ near $x_1 = -2$ and $x_1 = 2$, they are about to interact and produce another pair of kink lines. These newly formed kink lines move apart (see also the corresponding 2-D diagram in Fig. 9.4). At $t = 20$, one kink line from each of first and second periods have come quite near and are seen close to $x_1 = 0$. They interact, produce a new pair of kink lines and move apart as seen at $t = 30$, where we find four distinct kink lines in $-4 < x_1 < 4$. This process continues, two pairs of kink lines about to interact are seen near $x_1 = -2$ and $x_1 = 2$ at $t = 40$. There are four distinct kink lines also at $t = 50$ and at $t = 60$. The two kink lines near $x_1 = 0$ at $t = 60$ are about to interact.

The interaction of kinks was first observed on a 2-D shock front experimentally in [54], cf. Fig. 6(b) and (c) and numerically in [32] shown in Fig. 9.4. We do observe the same phenomenon here, except that kinks are replaced by kink lines. A detailed theoretical analysis of this phenomenon on a 2-D nonlinear wavefront is presented in [7] (see Sects. 8.5–8.7 for a brief description).

Equations (8.18) and (8.19) of WNLRT form a hyperbolic system of three conservation laws in (ξ, t)-plane when $m > 1$. This system is not degenerate and there are no source terms. The four elementary shapes corresponding to the two genuinely nonlinear characteristic fields consists of two shocks and two centred rarefaction waves in (ξ, t)-plane. Interaction of the elementary shapes in (x_1, x_2)-plane was studied with the help of interactions of elementary waves in (ξ, t)-plane. These interactions can be of finite or infinite duration in time. Two kinks of a different characteristic family on Ω_t approach interact and then produce another pair of kinks, which move apart.

For $M > 1$, the KCL-based 2-D SRT equations form a hyperbolic system of balance laws due to appearance of source terms in (8.26). However, as we have mentioned in Sect. 8.8 the source terms, which affect the solution on larger space and time scales, do not have any effect on the interaction of two kinks on a shock front, which takes place instantaneously. They also have only a small effect on the motion of kinks over a short time before and after the interaction. Thus, the nature of interaction of two kinks on a shock front as seen in Fig. 9.3 is same as that of interaction of kinks on a nonlinear wavefront theoretically predicted in Sect. 8.7. Interactions of two parallel kink lines on a 3-D shock front will be similar to that of the two kinks on a 2-D shock front. We clearly observe this in Fig. 10.3 for the interaction of a pair of parallel kink lines. However, the interaction of oblique kink lines (not discussed here) will be quite different.

Remark 10.4.1 Comparing the shock fronts at different times in Fig. 10.3, with the corresponding nonlinear wavefronts (see Fig. 8 in [1]) at the same time, we notice that the two fronts differ in their shapes. Let us compare the graphs of $M_{\max}(t)$ and $M_{\min}(t)$ in Fig. 10.4 with those of $m_{\max}(t)$ and $m_{\min}(t)$ in Fig. 11(a) of [1]. We find that $M_{\max}(t)$ and $M_{\min}(t)$ values on the shock front decay very rapidly compared to those on the nonlinear wavefront. As a result, the waves on the nonlinear wavefront move faster and the interaction of kink lines takes place more frequently. Graphs of $m_{\max}(t)$ and $m_{\min}(t)$ in Fig. 11a of [1] oscillate quite fast. Since sudden increases in the values of $M_{\max}(t)$ and $M_{\min}(t)$ (and similarly those of $m_{\max}(t)$ and $m_{\min}(t)$) correspond to interactions of kink lines, we notice that kink lines on the nonlinear wavefront interact more frequently than those on a shock front.

10.4.2 A Shock Front Starting from an Axisymmetric Shape

Next, we consider an initial shock front with an axisymmetric, oscillatory and radially decaying shape given by

$$\Omega_0 : x_3 = \kappa \cos(\alpha r) e^{-\beta r}, \tag{10.72}$$

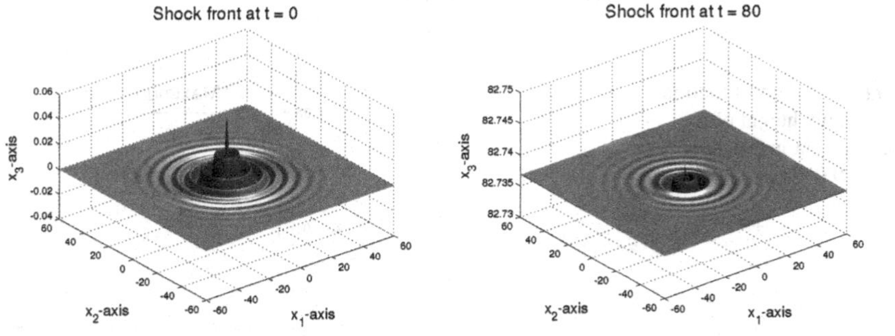

Fig. 10.5 The evolution of a shock front Ω_t starting from a smooth pulse $x_3 = 0.05\cos(r)e^{-0.15r}$

where $r = \sqrt{x_1^2 + x_2^2}$. We present here as yet another instance of corrugational stability. The initial shock front Ω_0 models a smooth perturbation of a planar front such that the amplitude of the perturbation decays to zero as $r \to \infty$. The parameters in (10.72) are taken as $\kappa = 0.05$, $\alpha = 1.0$, $\beta = 0.15$. The initial Mach number has a constant value $M_0 = 1.2$ and $\mathcal{V}_0 = 0.2$ everywhere on the shock front. In Fig. 10.5, we give the surface plots of the initial shock front Ω_0 and the shock front Ω_t at time $t = 80$. It can be noted that the front Ω_t moves up in the x_3-direction. At $t = 0$, there is an axisymmetric elevation (=0.5) at the origin $r = 0$ and this central elevation decays fast. The smooth shape Ω_0 at $t = 0$ develops later a number of circular kink lines, however these kink lines have almost disappeared at $t = 80$ since the height of the shock front has become quite small at this time. The elevations and depressions on the front diminish, leading to the reduction in height. We compute the maximum height $h(t)$ defined by

$$h(t) := x_{3\max}(t) - x_{3\min}(t), . \tag{10.73}$$

In Fig. 10.6, we give the plot of h versus t, which clearly shows that height reduces with time. The initial maximum height is $h(0) = 0.08$, whereas at $t = 80$ it is $h(80) = 0.002917$, which corresponds to a 96.35% reduction in the initial height. It is, therefore, very easy to see that the shock front tends to become planar, with its height decreasing to zero. In this decrease of the height of $h(t)$, a significant contribution comes from geometrical spreading of the cylindrical wave.

Now, we show that the normal velocity M of the shock and the gradient $\mathcal{V}$ of the gas density at the shock tend to become constant as the computational time increases. As before, let us denote by $M_{\max}(t)$ and $M_{\min}(t)$, the maximum and minimum of M taken over (ξ_1, ξ_2) at any time t. In an analogous manner, we define $\mathcal{V}_{\max}(t)$ and $\mathcal{V}_{\min}(t)$. In Fig. 10.7a we plot the distribution of $M_{\max}(t)$, $M_{\min}(t)$ and in Fig. 10.7b we plot $M_{\max}(t) - M_{\min}(t)$ with respect to time from $t = 0$ to $t = 80$. It can be seen that both $M_{\max}(t)$ and $M_{\min}(t)$ decay to one with time and as a result, the difference $M_{\max}(t) - M_{\min}(t)$ tends to zero asymptotically. We have also calculated $\mathcal{V}_{\max}(t)$, $\mathcal{V}_{\min}(t)$ and the difference $\mathcal{V}_{\max}(t) - \mathcal{V}_{\min}(t)$ and we have found that both $\mathcal{V}_{\max}(t)$ and $\mathcal{V}_{\min}(t)$ approach zero as time increases.

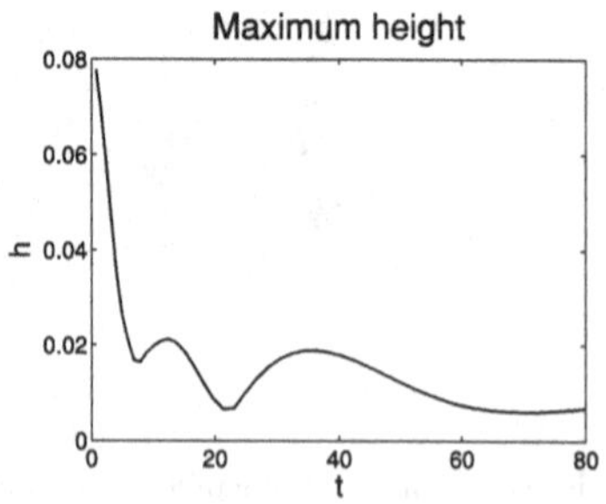

Fig. 10.6 Time variation maximum height of the shock front given initially by $x_3 = 0.05\cos(r)e^{-0.15r}$ eps

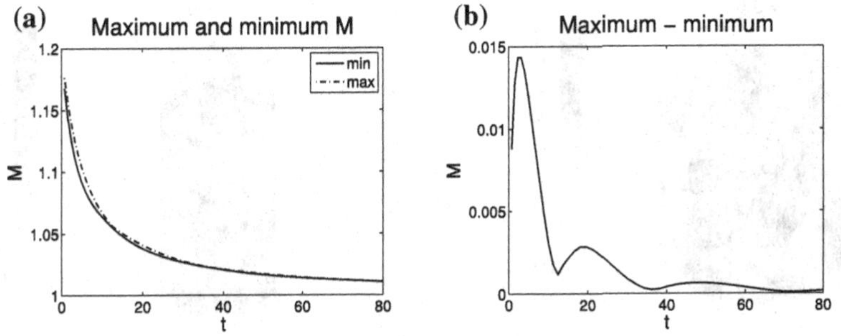

Fig. 10.7 For an initial shock front $x_3 = 0.05\cos(r)e^{-0.15r}$ (a): variation of $\mathcal{V}_{\max}(t)$ and $\mathcal{V}_{\min}(t)$ with time from $t = 0$ to $t = 80$. (b): the difference $\mathcal{V}_{\max}(t) - \mathcal{V}_{\min}(t)$ tends to zero as $t \to \infty$

10.4.3 Converging Shock Front Initially in the Shape of a Circular Cylinder

We present now the results of simulation of a cylindrically converging shock front. Even though this is basically a 2-D shock propagation problem, we intend to study it with our 3-D numerical method.

The initial geometry of the front is a portion of a circular cylinder of radius two units, i.e.

$$\Omega_0:\ x_1^2 + x_2^2 = 4,\ -\frac{\pi}{2} \le x_3 \le \frac{\pi}{2}. \tag{10.74}$$

Initially, the ray coordinates (ξ_1, ξ_2, t) are chosen as $\xi_1 = x_3$, $\xi_2 = \theta$ and $t = 0$, where θ is the azimuthal angle. Therefore, the initial shock front Ω_0 given in (10.74), can be expressed in a parametric form as follows:

$$\Omega_0:\ x_1 = 2\cos\xi_2,\ x_2 = 2\sin\xi_2,\ x_3 = \xi_1,\ -\frac{\pi}{2} \le \xi_1 \le \frac{\pi}{2},\ 0 \le \xi_2 \le 2\pi. \tag{10.75}$$

We have imposed periodic boundary conditions at $\xi_2 = 0$ and $\xi_2 = 2\pi$ and extrapolation boundary conditions at $\xi_1 = \pm\pi/2$.

As a result of the particular choice of the ray coordinates (ξ_1, ξ_2, t), the unit normal to Ω_0 given by $\boldsymbol{n}_0 = (-\cos\xi_2, \sin\xi_2, 0)$, points inward and hence the front converges. If the initial velocity is given a uniform distribution on Ω_0 as in the previous problems, the front Ω_t at any successive time t will remain a circular cylinder with no interesting geometrical features. Our aim here is to study the stability of a focusing shock front to perturbations. Hence, the initial distribution of the normal velocity M is given as a small perturbation of a constant value, i.e.

$$M_0(\xi_1, \xi_2) = 1.2 + \alpha\cos(\nu\xi_2) \tag{10.76}$$

with $\alpha = 0.05$ and $\nu = 8$. Here also we have taken $\mathcal{V}_0 = 0.2$.

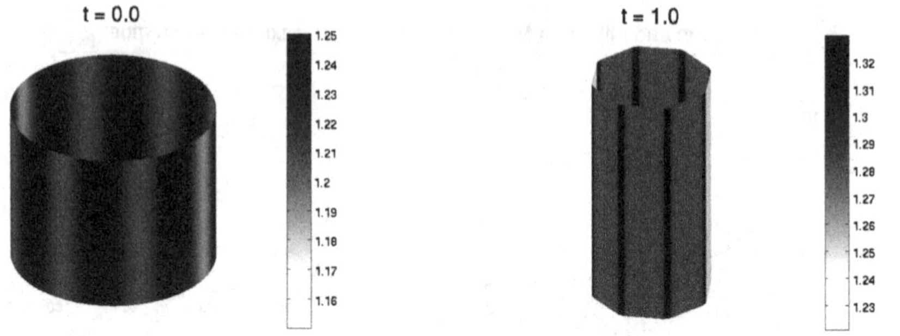

Fig. 10.8 Cylindrically converging shock front. On the left the initial front and on the right the front at time $t = 1.0$. The greyscale-bar on the right-hand side indicates the intensity of the normal velocity M

In Fig. 10.8 we give the plots of the initial shock front Ω_0 and the shock front Ω_t at time $t = 1.0$. The shock fronts are coloured using the variation of the normal velocity M, with the greyscale-bar on the right indicating the values of M. Note that the initial normal velocity M_0 given by (10.76) has a periodic variation with a maximum value 1.25 and a minimum value 1.05. Those portions of the front where M_0 has maximum value moves inwards faster and it results in a distortion of the circular shape of Ω_0. From the shock front at $t = 1$ in Fig. 10.9, it can be observed that 16 fully developed vertical kink lines are formed on the shock front and kink lines have not yet interacted. Clearly, the number of plane sides and kink lines on Ω_t at various times will depend on the value of parameter ν in (10.76). The shock front assumes the shape of a polygonal cylinder and as time progresses the kink lines on the shock front interact, with the formation of new kink lines and this process repeats. It has to be remarked that our numerical results are well in accordance with the experimental results of [55], where the authors have reported that a converging shock front with a small perturbation assumes the shape of a polygon. This test problem gives yet another evidence for the efficacy of SRT to produce very clear and physically realistic geometrical features (Fig. 10.8).

An examination of the shock front at $t = 1.0$ in Fig. 10.8 and the corresponding nonlinear wavefront in [1] shows that both the fronts have analogous geometry. We also refer the reader to [1] for more details on the transient geometries between the initial and final configurations and a quantification of the focusing process. Nevertheless, in order to point out the important difference in the case of a shock front, in Fig. 10.9 we give the successive cross-sections (say by $x_3 = 0$ plane) of the shock fronts and the corresponding nonlinear wavefronts at times $t = 0, 0.1, 0.2, \ldots, 1.0$. Note that initially the normal speeds m and M of both the fronts have the same initial values. These speeds increase due to convergence of the fronts, however, in the case of a shock front there is also a decay (due to the second terms in (7.20) and (7.21)) or according to the right-hand side terms in (5.56) and (5.57) in the shock speed due to interaction of the shock with the nonlinear wavefronts. Therefore, as time increases, the shock front lags behind the nonlinear wavefront as seen in Fig. 10.9.

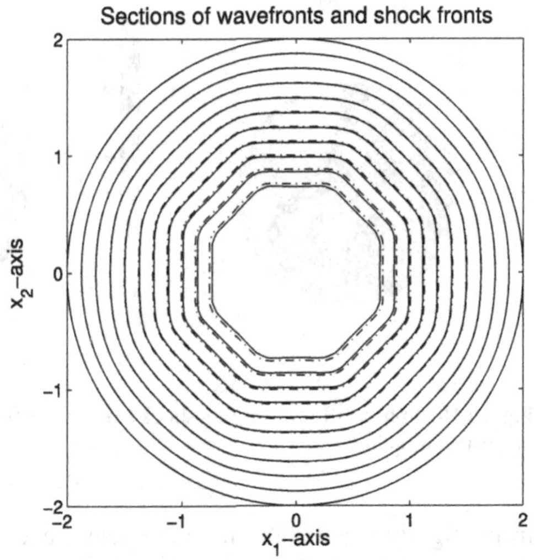

Fig. 10.9 Cross-sections of the shock fronts (broken lines) and the nonlinear wavefront (continuous lines) at $t = 0, 0.1, 0.2, \ldots, 1.0$ by $x_3 = 0$ plane

10.4.4 Spherically Converging Shock Front

In this section, we consider the propagation of a spherically converging shock front. The initial geometry of the shock front is a sphere of radius 2 units. For the ray coordinates (ξ_1, ξ_2, t), at $t = 0$ we choose $\xi_1 = \pi - \phi$ and $\xi_2 = \theta$, where θ is the azimuthal angle and ϕ is the polar angle. Therefore, the parametric representation of the initial shock front Ω_0 is

$$x_1 = 2\sin\xi_1\cos\xi_2, \quad x_2 = 2\sin\xi_1\sin\xi_2, \quad x_3 = -2\cos\xi_1. \tag{10.77}$$

In order to avoid the singularities at $\phi = 0$ and $\phi = \pi$, we remove these points. Therefore, our computational domain is $[\pi/15, 14\pi/15] \times [0, 2\pi]$. As in the previous problem, we choose the initial velocity distribution as a small perturbation of a constant value, i.e.

$$M_0(\xi_1, \xi_2) = 1.2 + \alpha\cos(\nu_1\xi_1)\cos(\nu_2\xi_2) \tag{10.78}$$

with $\alpha = 0.05$, $\nu_1 = 4$, $\nu_2 = 8$. Here also we have taken $\mathcal{V}_0 = 0.2$.

The 3-D plots of the initial shock front and the one at time $t = 0.85$ are given in Fig. 10.10. The shock fronts are coloured using the variation of the normal velocity M, with the greyscale-bar on the right indicating the values of M. It can be observed from the figure that as the front starts focusing, it develops several kink curves and its spherical shape gets distorted, with the formation of facets. The final shape of the shock front is almost a polyhedron.

In this problem also, we have observed a qualitatively similar behaviour of a shock front and nonlinear wavefront in [1]. The major important difference is that the shock

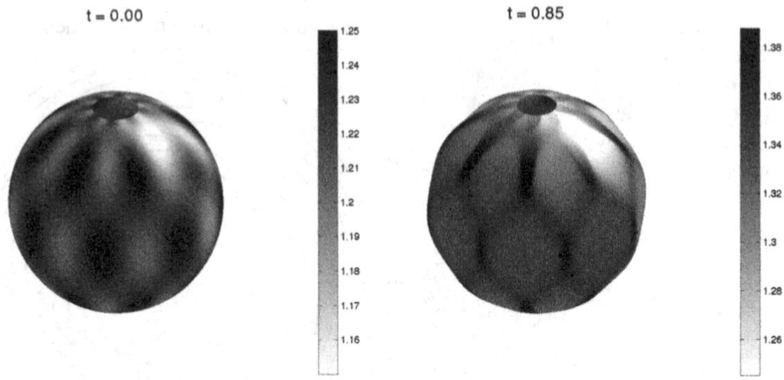

Fig. 10.10 Spherically converging shock front at $t = 0.85$. The greyscale bar on the right represents the distribution of M

front slightly lags behind the nonlinear wave due to the effect on the shock of the flow behind the shock, as explained in the case of a cylindrically converging shock.

Remark 10.4.2 It has to be emphasized that we can continue the computations further, however, the results would not be physically realistic as the weak shock ray theory breaks down for larger values of $M - 1$.

Remark 10.4.3 As in the case nonlinear wavefront in [1], we notice that cylindrically and spherically converging shocks take the polygonal and polyhedral shapes, respectively. These two shapes are stable configurations for the two cases. The stability here does not mean that once one of these configurations is formed, the shock front remains in this configuration for all time. It may change again into another such configuration.

10.5 Some More Comparison of Results of 3-D WNLRT and 3-D SRT

By comparing the results of numerical experiments presented above with those of 3-D WNLRT reported in [1], we infer that the geometrical shapes of nonlinear wavefronts and shock fronts are qualitatively similar. However, looking at these shapes alone, it is not possible say much about the limiting results as $t \to \infty$.

Observation 1: When the initial shape is periodic in x_1 and x_2 given by (10.71), we look at the behaviour of a nonlinear wavefront in [1] and that of a shock front in Sect. 10.4.1 for $t \to \infty$ and find that both are corrugationally stable. This means that both these fronts tend to become planar after a long time, which in turn shows that the mean curvature Ω approaches zero. Taking differential forms (7.6) and (7.7) of (10.6) and letting $\Omega \to 0$, i.e. the mean curvature approaching zero yields

$$m(\xi_1, \xi_2, t) \to \ constant, \ \ as \ \ t \to \infty \quad along\, a\, ray. \tag{10.79}$$

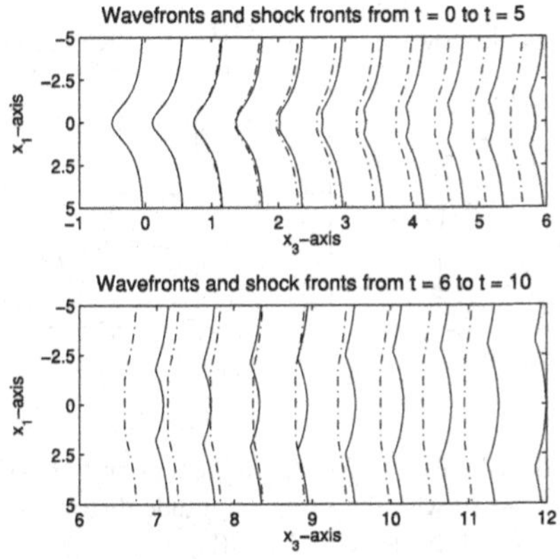

Fig. 10.11 Comparison of the results by 3-D WNLRT and 3-D SRT. Sections of the fronts in $x_2 = 0$ for fronts initially given by (10.80). Figure on the top: from $t = 0.0$ to $t = 5.0$ and bottom: from $t = 6.0$ to $t = 10.0$. The solid lines represent successive positions of the nonlinear wavefront obtained by 3-D WNLRT and dotted lines are those of the shock front from 3-D SRT. The kinks can be noticed on both the fronts. The nonlinear wavefront overtakes the shock and its central portion bulges out so that the kinks on it are very prominent

In order that the nonlinear wavefront be corrugationally stable, this constant must be the same along all rays. It has been observed in Fig. 11 of [1] that even though the maximum and minimum of the front velocity m of a nonlinear wavefront heavily oscillate about their initial values, they finally approach their initial mean values as $t \to \infty$ as seen also in Fig. 10.12a. It is, therefore, very interesting to note that the numerical computation suggests the constant in (10.79) is m_0 for the case when m_0 is constant everywhere on the initial wavefront Ω_0 (Fig. 10.11).

We now look at the long-term behaviour of the perturbations on a 3-D corrugationally stable plane shock front. First, we note that the numerical results in Fig. 10.4 (seen also in Fig. 10.12b) show that both $M_{\max}$ and $M_{\min}$ decrease to one with increasing time. Therefore, (as observed also in Sects. 10.4.1 and 10.4.2) we again see that $M \to 1$ as $t \to \infty$. The results also show that the gradient $\mathcal{V}$ of the pressure decays to zero as $t \to \infty$. Analogous results were reported also in the 2-D case in [32]. Thus, we can see a major quantitative difference in the decay of the shock amplitude from the corresponding results of 3-D WNLRT.

Observation 2: In order to do some more comparison of results for a nonlinear wavefront and shock front, we choose the initial geometry of the nonlinear wavefront and shock front to be an non-axisymmetric dip given by

$$\Omega_0\colon\ x_3 = \frac{-\kappa}{1 + \frac{x_1^2}{\alpha^2} + \frac{x_2^2}{\beta^2}}, \tag{10.80}$$

where the parameter values are set to $\kappa = 0.5, \alpha = 1.5, \beta = 3$. We take the same

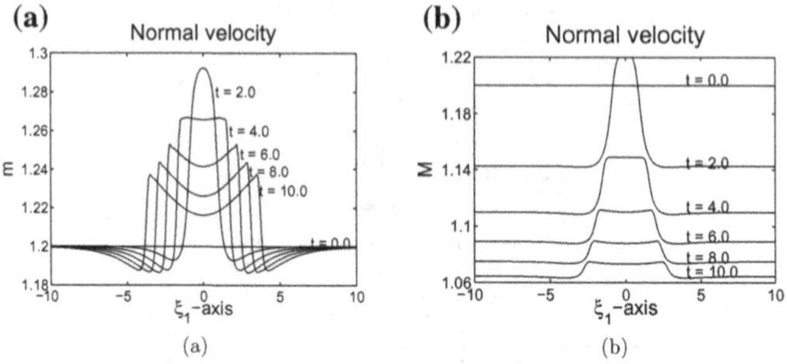

Fig. 10.12 Comparison of distributions of Mach numbers m and M on the fronts (initially given by (10.80)) at times 0, 2, 4, 6, 8, 10 with the same amplitude distribution 1.2 on the initial fronts. **a** Results obtained by 3-D WNLRT. **b** Those of 3-D SRT

initial values of M and m, both equal to 1.2, where m and M are given by (5.50). For SRT we take $\mathcal{V} = 0.2$. The computations are done with 3-D WNLRT and 3-D SRT up to a time $t = 10$. A comparison of the results obtained is presented in Fig. 10.11, where we have plotted the successive wavefronts and shock fronts in the section $x_2 = 0$ from $t = 0$ to $t = 10$ in a time step of 0.5. In the figure, the solid lines represent the successive nonlinear wavefronts and dotted lines are the shock fronts. The figure clearly shows that from time $t = 2.0$ onwards, the nonlinear wavefront overtakes the corresponding shock front. We also notice that the central portion of the nonlinear wavefront bulges out and the two kinks move apart faster than those on the shock front.

Observation 3: We also present in Fig. 10.12 the graphs of the normal velocity m of the nonlinear wavefront and M of that of the shock front. In the figure we have plotted both m and M at times 0, 2, 4, 6, 8, 10 with the same constant initial values 1.2 on the respective fronts. The Mach number at the centre of the fronts initially rises considerably in both the cases but it becomes constant on the central disc at time $t = 4$. As the rays starts diverging from the bulged central portion, cf. Figure 10.12a, we can see that m reduces at the central portion from $t = 6$ onwards but finally it tends to 1.2 on all parts of the nonlinear wavefront. However, the shock Mach number (and hence the shock amplitude) remains constant after $t > 6$ on the central disc and as seen in all previous cases with $\mathcal{V}_0 > 0$, decreases to 0 with time on all parts of the shock front governed by 3-D SRT.

10.6 Appendices

Appendix 1

Non-zero elements of the matrices A, $B^{(1)}$ and $B^{(2)}$ in (10.23):

$$
\begin{aligned}
a_{11} &= g_1, \quad a_{22} = g_1, \quad a_{33} = g_2, \quad a_{44} = g_2,\\
a_{51} &= -\frac{1}{u_3} g_1 g_2 n_2 \cot\chi, \quad a_{52} = \frac{1}{u_3} g_1 g_2 n_1 \cot\chi,\\
a_{53} &= \frac{1}{v_3} g_1 g_2 n_2 \cot\chi, \quad a_{54} = -\frac{1}{v_3} g_1 g_2 n_1 \cot\chi,\\
a_{55} &= \frac{2m}{m-1} g_1 g_2, \quad a_{56} = g_2, \quad a_{57} = g_1, \quad a_{66} = 1, \quad a_{77} = 1.\\
b^{(1)}_{11} &= -\frac{m}{u_3}(u_1 u_2 + n_1 n_2)\cot\chi, \; b^{(1)}_{12} = \frac{m}{u_3}(u_1^2 + n_1^2 - 1)\cot\chi,\\
b^{(1)}_{13} &= \frac{m}{v_3 \sin\chi}(u_2 v_1 + n_1 n_2 \cos\chi), \; b^{(1)}_{14} = -\frac{m}{v_3 \sin\chi}(u_1 v_1 + (n_1^2 - 1)\cos\chi), \; b^{(1)}_{15} = -n_1,\\
b^{(1)}_{21} &= -\frac{m}{u_3}(u_2^2 + n_2^2 - 1)\cot\chi, \; b^{(1)}_{22} = \frac{m}{u_3}(u_1 u_2 + n_1 n_2)\cot\chi,\\
b^{(1)}_{23} &= \frac{m}{v_3 \sin\chi}(u_2 v_2 + (n_2^2 - 1)\cos\chi), \; b^{(1)}_{24} = -\frac{m}{v_3 \sin\chi}(u_1 v_2 + n_1 n_2 \cos\chi), \; b^{(1)}_{25} = -n_2.\\
b^{(1)}_{61} &= -\frac{m}{u_3 \sin\chi}(v_2 - u_2\cos\chi), \; b^{(1)}_{62} = \frac{m}{u_3 \sin\chi}(v_1 - u_1\cos\chi).\\
b^{(2)}_{31} &= -\frac{m}{u_3 \sin\chi}(u_1 v_2 + n_1 n_2 \cos\chi), \; b^{(2)}_{32} = \frac{m}{u_3 \sin\chi}(u_1 v_1 + (n_1^2 - 1)\cos\chi),\\
b^{(2)}_{33} &= \frac{m}{v_3}(v_1 v_2 + n_1 n_2)\cot\chi, \; b^{(2)}_{34} = -\frac{m}{v_3}(v_1^2 + n_1^2 - 1)\cot\chi, \; b^{(2)}_{35} = -n_1.\\
b^{(2)}_{41} &= -\frac{m}{u_3 \sin\chi}(u_2 v_2 + (n_2^2 - 1)\cos\chi), \; b^{(2)}_{42} = \frac{m}{u_3 \sin\chi}(u_2 v_1 + n_1 n_2 \cos\chi),\\
b^{(2)}_{43} &= \frac{m}{v_3}(v_2^2 + n_2^2 - 1)\cot\chi, \; b^{(2)}_{44} = -\frac{m}{v_3}(v_1 v_2 + n_1 n_2)\cot\chi, \; b^{2}_{45} = -n_2.\\
b^{(2)}_{73} &= \frac{m}{v_3 \sin\chi}(u_2 - v_2\cos\chi), \; b^{(2)}_{74} = -\frac{m}{v_3 \sin\chi}(u_1 - v_1\cos\chi).
\end{aligned}
$$

Appendix 2

Non-zero elements of the matrices $\tilde{A}$, $\tilde{B}^{(1)}$, $\tilde{B}^{(2)}$ and $\tilde{C}$ of (10.40):

$$
\begin{aligned}
\tilde{a}_{11} &= G_1, \quad \tilde{a}_{22} = G_1, \quad \tilde{a}_{33} = G_2, \quad \tilde{a}_{44} = G_2,\\
\tilde{a}_{51} &= -\frac{1}{U_3} G_1 G_2 N_2 \cot\Psi, \; \tilde{a}_{52} = \frac{1}{U_3} G_1 G_2 N_1 \cot\Psi,\\
\tilde{a}_{53} &= \frac{1}{V_3} G_1 G_2 N_2 \cot\Psi, \; \tilde{a}_{54} = -\frac{1}{V_3} G_1 G_2 N_1 \cot\Psi,\\
\tilde{a}_{55} &= \frac{2M}{M-1} G_1 G_2, \; \tilde{a}_{56} = G_2, \; \tilde{a}_{57} = G_1, \tilde{a}_{66} = \tilde{a}_{77} = 1,\\
\tilde{a}_{81} &= -\frac{1}{U_3} G_1 G_2 N_2 \cot\Psi, \; \tilde{a}_{82} = \frac{1}{U_3} G_1 G_2 N_1 \cot\Psi, \; \tilde{a}_{83} = \frac{1}{V_3} G_1 G_2 N_2 \cot\Psi,\\
\tilde{a}_{84} &= -\frac{1}{V_3} G_1 G_2 N_1 \cot\Psi, \; \tilde{a}_{85} = 2 G_1 G_2, \; \tilde{a}_{86} = G_2, \; \tilde{a}_{87} = G_1, \; \tilde{a}_{88} = \frac{2 G_1 G_2}{\mathcal{V}}.
\end{aligned}
$$

$$\tilde{b}_{11}^{(1)} = -\frac{M}{U_3}(U_1U_2 + N_1N_2)\cot\Psi,\ \tilde{b}_{12}^{(1)} = \frac{M}{U_3}(U_1^2 + N_1^2 - 1)\cot\Psi,$$

$$\tilde{b}_{13}^{(1)} = \frac{M}{V_3\sin\Psi}(U_2V_1 + N_1N_2\cos\Psi),\ \tilde{b}_{14}^{(1)} = -\frac{M}{V_3\sin\Psi}(U_1V_1 + (N_1^2 - 1)\cos\Psi),\ \tilde{b}_{15}^{(1)} = -N_1.$$

$$\tilde{b}_{21}^{(1)} = -\frac{M}{U_3}(U_2^2 + N_2^2 - 1)\cot\Psi,\ \tilde{b}_{22}^{(1)} = \frac{M}{U_3}(U_1U_2 + N_1N_2)\cot\Psi,$$

$$\tilde{b}_{23}^{(1)} = \frac{M}{V_3\sin\Psi}(U_2V_2 + (N_2^2 - 1)\cos\Psi),\ \tilde{b}_{24}^{(1)} = -\frac{M}{V_3\sin\Psi}(U_1V_2 + N_1N_2\cos\Psi),\ \tilde{b}_{25}^{(1)} = -N_2.$$

$$\tilde{b}_{31}^{(2)} = -\frac{M}{U_3\sin\Psi}(U_1V_2 + N_1N_2\cos\Psi),\ \tilde{b}_{32}^{(2)} = \frac{M}{U_3\sin\Psi}(U_1V_1 + (N_1^2 - 1)\cos\Psi),$$

$$\tilde{b}_{33}^{(2)} = \frac{M}{V_3}(V_1V_2 + N_1N_2)\cot\Psi,\ \tilde{b}_{34}^{(2)} = -\frac{M}{V_3}(V_1^2 + N_1^2 - 1)\cot\Psi,\ \tilde{b}_{35}^{(2)} = -N_1.$$

$$\tilde{b}_{41}^{(2)} = -\frac{M}{U_3\sin\Psi}(U_2V_2 + (N_2^2 - 1)\cos\Psi),\ \tilde{b}_{42}^{(2)} = \frac{M}{U_3\sin\Psi}(U_2V_1 + N_1N_2\cos\Psi),$$

$$\tilde{b}_{43}^{(2)} = \frac{M}{V_3}(V_2^2 + N_2^2 - 1)\cot\Psi,\ \tilde{b}_{44}^{(2)} = -\frac{M}{V_3}(V_1V_2 + N_1N_2)\cot\Psi,\ \tilde{b}_{45}^{2} = -N_2.$$

$$\tilde{b}_{61}^{(1)} = -\frac{M}{U_3\sin\Psi}(V_2 - U_2\cos\Psi),\ \tilde{b}_{62}^{(1)} = \frac{M}{U_3\sin\Psi}(V_1 - U_1\cos\Psi).$$

$$\tilde{b}_{73}^{(2)} = \frac{M}{V_3\sin\chi}(U_2 - V_2\cos\chi),\ \tilde{b}_{74}^{(2)} = -\frac{M}{V_3\sin\Psi}(U_1 - V_1\cos\Psi).$$

$$\tilde{c}_7 = -\frac{2M}{M-1}G_1G_2\mathcal{V},\ \tilde{c}_8 = -2MG_1G_2\mathcal{V}.$$

Bibliography

1. K.R. Arun, A numerical scheme for three-dimensional front propagation and control of Jordan mode. SIAM J. Sci. Comput. **34**, B148–B178 (2012)
2. K.R. Arun, M. Lukacova-Medvidova, P. Prasad, S.V. Raghurama, Rao, An application of 3-D kinematical conservation laws: propagation of a three dimensional wavefront. SIAM J. Appl. Math. **70**, 2604–2626 (2010)
3. K.R. Arun, P. Prasad, 3-D kinematical conservation laws (KCL): equations of evolution of a surface. Wave Motion **46**, 293–311 (2009)
4. K.R. Arun, P. Prasad, Eigenvalues of kinematical conservation laws (KCL) based 3-D weakly nonlinear ray theory (WNLRT). Appl. Math. Comput. **217**, 2285–2288 (2010)
5. K.R. Arun, P. Prasad, *Propagation of a three-dimensional weak shock front using kinematical conservation laws*, when the paper was ready, authors first wrote to JFM editor about this paper on September 1, 2010. arXiv:1709.06791 [math.AP] (2017)
6. S. Baskar, P. Prasad, Kinematical conservation laws applied to study geometrical shapes of a solitary wave, in *Wind over Waves II: Forecasting and Fundamentals*, ed. by S. Sajjadi, J. Hunt (Horwood Publishing Ltd, 2003), pp. 189–200
7. S. Baskar, P. Prasad, Riemann problem for kinematical conservation laws and geometrical features of nonlinear wavefronts. IMA J. Appl. Math. **69**, 391–420 (2004)
8. S. Baskar, P. Prasad, Propagation of curved shock fronts using shock ray theory and comparison with other theories. J. Fluid Mech. **523**, 171–198 (2005)
9. S. Baskar, P. Prasad, Formulation of the problem of sonic boom by a maneuvering aerofoil as a one parameter family of Cauchy problems. Proc. Indian Acad. Sci. (Math. Sci.) **116**, 97–119 (2006)
10. P.L. Bhatnagar, *Nonlinear Waves in One-dimensional Dispersive Systems, Oxford Mathematical Monograph Series* (Oxford University Press, New York, 1979)
11. R.N. Buchal, J.B. Keller, Boundary layer problems in diffraction theory. Commun. Pure Appl. Math. **13**, 85–144 (1960)
12. Y. Choquet-Bruhat, Ondes Asymptotiques et Approchees pour des Systemes d'Equations aux Derivées Partielles non Linéaires. J. Math. Pures et Appl. **48**, 117–158 (1969)
13. R. Courant, D. Hilbert, *Methods of Mathematical Physics: Partial Differential Equations*, vol. 2 (Interscience Publishers, New York, 1962)
14. R. Courant, K.O. Friedrichs, *Supersonic Flow and Shock Waves* (Interscience Publishers, New York, 1948)
15. R. Courant, F. John, *Introduction to Calculus and Analysis*, vol. II (Wiley, New York, 1974)

P. Prasad, *Propagation of Multidimensional Nonlinear Waves and Kinematical Conservation Laws*, Infosys Science Foundation Series,
https://doi.org/10.1007/978-981-10-7581-0

16. C.M. Dafermos, *Hyperbolic Conservation Laws in Continuum Physics* (Springer, Berlin, 2016)
17. L.C. Evans, *Partial Differential Equations*, vol. 19, Graduate Studies in Mathematics (American Mathematical Society, 1999)
18. C.R. Evans, J.F. Hawley, Simulation of magnetohydrodynamic flows-a constrained transport method. Astrophys. J. **332**, 659–677 (1988)
19. I.M. Gel'fand, Some problems in the theory of quasilinear equations. Amer. Math. Soc. Translations, Ser. 2 **29**, 295–381 (1963)
20. M. Giles, P. Prasad, R. Ravindran, *Conservation form of equations of three dimensional front propagation* (Technical Report, Department of Mathematics, Indian Institute of Science, Bangalore, 1995)
21. E. Godlewski, P.A. Raviart, *Numerical Approximation of Hyperbolic Systems of Conservation Laws* (Springer, Berlin, 1996)
22. M.A. Grinfel'd, Ray method for calculating the wavefront intensity in nonlinear elastic material. PMM J. Appl. Math. Mech. **42**, 958–77 (1978)
23. J.K. Hunter, J.B. Keller, Weakly nonlinear high frequency waves. Commun. Pure Appl. Math. **36**, 547–69 (1983)
24. O. Inoue, T. Sakai, M. Nishida, Focusing shock wave generated by an accelerating projectile. Fluid Dyn. Res. **21**, 403–416 (1997)
25. G.-S. Jiang, C.-W. Shu, Efficient implementation of weighted ENO schemes. J. Comput. Phys. **126**, 202–228 (1996)
26. N.K.R. Kevlahan, The propagation of weak shocks in non-uniform flow. J. Fluid Mech. **327**, 167–97 (1996)
27. A. Kurganov, E. Tadmor, New high-resolution central schemes for nonlinear conservation laws and convection-diffusion equations. J. Comput. Phys. **160**, 241–282 (2000)
28. P.D. Lax, Hyperbolic system of conservation laws II. Commun. Pure Appl. Math. **10**, 537–66 (1957)
29. P.D. Lax, *On hyperbolic partial differential equations* (Stanford University, Stanford, 1963)
30. D. Ludwig, Uniform asymptotic expansions at a caustic. Commun. Pure Appl. Math. **19**, 215–50 (1966)
31. V.P. Maslov, Propagation of shock waves in an isentropic non-viscous gas. J. Sov. Math. **13**, 119–163 (1980). Russian publication 1978
32. A. Monica, P. Prasad, Propagation of a curved weak shock. J. Fluid Mech. **434**, 119–151 (2001)
33. K.W. Morton, P. Prasad, R. Ravindran, *Conservation forms of nonlinear ray equations*, Technical Report No.2, Department of Mathematics, Indian Institute of Science, Bangalore, 1992
34. D.F. Parker, Nonlinearity, relaxation and diffusion in acoustic and ultrasonics. J. Fluid Mech. **39**, 793–815 (1969)
35. D.F. Parker, An asymptotic theory for oscillatory nonlinear signals. J. Inst. Math. Appl. **7**, 92–110 (1971)
36. P. Prasad, *Introduction to hyperbolic PDE and nonlinear waves* (Lecture Note, Department of Applied Mathematics, Indian Institute of Science, 1977)
37. P. Prasad, Kinematics of a multi-dimensional shock of arbitrary strength in an ideal gas. Acta Mechanica **45**, 163–76 (1982)
38. P. Prasad, *Propagation of a Curved Shock and Nonlinear Ray Theory, Pitman Research Notes in Mathematics Series*, vol. 292 (Longman, Essex, 1993)
39. P. Prasad, 1. A nonlinear ray theory, Berichte der Arbeitsgruppe Technomathematik, Universitaet Kaiserslautern, **101**, 1993; 2. On the lemma on bicharacteristics, Appendix in this
40. P. Prasad, A nonlinear ray theory. Wave Motion **20**, 21–31 (1994)
41. P. Prasad, An asymptotic derivation of weakly nonlinear ray theory. Proc. Indian Acad. Sci. (Math. Sci.) **110**, 431–447 (2000)
42. P. Prasad, *Nonlinear Hyperbolic Waves in Multi-dimensions*, vol. 121 (Monographs and Surveys in Pure and Applied Mathematics (Chapman and Hall/CRC, Boca Raton, 2001)
43. P. Prasad, Ray theories for hyperbolic waves, kinematical conservation laws and applications. Indian J. Pure Appl. Math. **38**, 467–490 (2007)

44. P. Prasad, Fermat's and Huygens' principles, and hyperbolic equations and their equivalence in wavefront construction. Neural Parallel Sci. Comput. **21**, 305–318 (2013)
45. P. Prasad, Ray equations of a weak shock in a hyperbolic system of conservation laws in multi-dimensions. Proc. Indian Acad. Sci. (Math. Sci.) **126**, 199–206 (2016)
46. P. Prasad, Kinematical conservation laws in a space of arbitrary dimensions. Indian J. Pure Appl. Math. **47**, 641–653 (2016)
47. P. Prasad, R. Ravindran, *Partial Differential Equations* (Wiley Eastern Ltd, 1985)
48. T.M. Ramanathan, *Huygens' Method of Construction of Weakly Nonlinear Wavefronts and Shock Fronts*, Ph.D. Thesis, Indian Institute of Science, Bangalore, 1985
49. R. Ravindran, P. Prasad, A new theory of shock dynamics, part I: analytical considerations. Appl. Math. Lett. **3**, 77–81 (1990)
50. R. Ravindran, P. Prasad, On infinite system of compatibility conditions along a shock ray. Q. J. Appl. Math. Mech. **46**, 131–140 (1993)
51. J. Smoller, *Shock Waves and Reaction Diffusion Equation*, 2nd edn. (Springer, Berlin, 1994)
52. R. Srinivasan, *A Mathematical Theory on the Propagation of Multi-dimensional Shock and Nonlinear Waves*, Ph.D. Thesis, Indian Institute of Science, Bangalore, 1987
53. C.-W. Shu, Total-variation-diminishing time discretizations. SIAM J. Sci. Stat. Comput. **9**, 1073–84 (1988)
54. B. Sturtevant, V.A. Kulkarni, The focusing of weak shock waves. J. Fluid Mech. **73**, 651–71 (1976)
55. K. Takayama, H. Kleine, H. Grönig, An experimental investigation of the stability of converging cylindrical shock waves in air. Exp. Fluids **5**, 315–322 (1987). ISSN 0723-4864
56. G.B. Whitham, *Linear and Nonlinear Waves* (John Wiley & Sons, New York, 1974)

Index

P. Prasad, *Propagation of Multidimensional Nonlinear Waves and Kinematical Conservation Laws*, Infosys Science Foundation Series,
https://doi.org/10.1007/978-981-10-7581-0

Printed in the United States
By Bookmasters